全国高等学校计算机教育研究会“十四五”规划教材

国家精品在线开放课程教材

高等教育国家级教学成果二等奖

清華大學 计算机系列教材

郑 莉　尹 刚　张 宇　编著

# Java语言程序设计实践教程

清華大學出版社
北 京

## 内容简介

本书是与《Java语言程序设计》(第3版)(ISBN 978-7-302-58165-9)配套的实践教程，能够配合主教材为读者提供学习指导，帮助读者提高编程实践能力。

本书首先给出一个简要的"学习指南"，其余各章与主教材内容相对应，每章包括"要点导读""实验指导"，以及与主教材每章习题配套的"习题解答"。"实验指导"包括"实验目的""实验任务""实验步骤"。"习题解答"给出了主教材中各章习题的参考答案。



**图书在版编目(CIP)数据**

Java语言程序设计实践教程/郑莉，尹刚，张宇编著．—北京：清华大学出版社，2023.6(2024.9重印)
清华大学计算机系列教材
ISBN 978-7-302-63315-0

Ⅰ.①J… Ⅱ.①郑… ②尹… ③张… Ⅲ.①JAVA语言－程序设计－高等学校－教材 Ⅳ.①TP312.8

中国国家版本馆CIP数据核字(2023)第060493号

**责任编辑**：谢　琛　薛　阳
**封面设计**：常雪影
**责任校对**：申晓焕
**责任印制**：宋　林

**出版发行**：清华大学出版社
　　**网　　址**：https://www.tup.com.cn，https://www.wqxuetang.com
　　**地　　址**：北京清华大学学研大厦A座　　**邮　　编**：100084
　　**社 总 机**：010-83470000　　**邮　　购**：010-62786544
　　**投稿与读者服务**：010-62776969，c-service@tup.tsinghua.edu.cn
　　**质量反馈**：010-62772015，zhiliang@tup.tsinghua.edu.cn
　　**课件下载**：https://www.tup.com.cn，010-83470236
**印 装 者**：三河市铭诚印务有限公司
**经　　销**：全国新华书店
**开　　本**：185mm×260mm　　**印　　张**：9.5　　**字　　数**：234千字
**版　　次**：2023年6月第1版　　**印　　次**：2024年9月第3次印刷
**定　　价**：39.00元

产品编号：078479-01

# 前 言

计算机程序设计是一门实践性很强的课程，因此仅通过阅读教科书或听课是不可能完全掌握的，学习程序设计的最重要环节就是实践。对于自学读者来说，更多一重困难，就是在学习和实践过程中缺乏指导。

学习程序设计的人，往往有这样的感觉：看书或听课时，对老师讲的和书上写的内容基本上能够理解，但是当需要自己编程时却又无从下手。相信每一个讲授程序设计课程的教师都有过这样的经历：有些问题，尽管在课上再三强调，反复举例，学生还是不能够完全理解，上机时更是错误百出。应该说，这是学习过程中的必然现象。

要想能够将书本上的知识变为自己所具有的能力，需要的是实践、实践、再实践。在实践环节中，起主导作用的是学习者自己，旁人是无法代劳的，也不能期望有什么一蹴而就的捷径。但是，由于学生在实践过程中不能随时随地得到指导，因此花费时间较多，总感觉程序设计课程作业负担太重，有的学生甚至因为花费四五个小时调不通一个简单的程序而失去学习兴趣。

本书是与《Java语言程序设计》(第3版)配套的实践教程，目的是为读者的学习提供一些指导，为提高读者的编程能力助一臂之力，使读者在实践的过程中少些曲折和彷徨，多些成功的乐趣。本书出版之前已经在清华大学“Java语言程序设计”课程中使用，取得了良好的教学效果。

本书首先给出一个简要的“学习指南”，其余各章与主教材《Java语言程序设计》(第3版)相对应，每章内容分为三部分：第一部分是要点导读，主要是为自学读者指明学习重点，建议学习方法；第二部分是实验指导，每章都有一个精心设计的实验，与《Java语言程序设计》(第3版)相应章内容配合，使读者在实践中达到对主教材内容的深入理解和熟练掌握，每个实验都包括“实验目的”“实验任务”“实验步骤”，实验的完整参考程序可从清华大学出版社网站上下载；第三部分是习题解答，给出了《Java语言程序设计》(第3版)各章习题的参考答案。每个题目可能有多种解法，这里仅给出一种参考解法。

本书中的习题解答和实验内容，不仅可以指导读者上机练习，也可以由教师选作为例题在课上演示，使教学内容更加丰富。如果读者没有足够的时间完成全部的习题和实验，可以将剩下的题解作为例题阅读，这也不失为一种好的选择。

本书由郑莉、尹刚、张宇共同编写，参加本书编写工作的还有黄帅、张力兮。

感谢读者选择使用本书，欢迎您对本书内容提出意见和建议，我们将不胜感激。作者的电子邮件地址：zhengli@tsinghua.edu.cn，来信标题请包含“Java book”。

作　者

2023年3月于清华大学

# 《Java 语言程序设计》(第 3 版)学习指南

《Java 语言程序设计》(第 3 版)是针对初学 Java 语言的读者编写的入门教材，预期的读者主要有这样几类：初学 Java 语言的自学读者、以此为 Java 课程教材或参考书的在校学生、Java 课程教师。这里首先给出学习本套教材的总体建议，在本书的后续各章中还会有详细的导读。

一、主教材《Java 语言程序设计》(第 3 版)的学习方法

对于主教材的学习，读者需要时刻牢记两点：面向对象的编程方法；查阅 Java API 手册，掌握主教材中出现的包、类、方法，并了解 API 中的其他包、类、方法。

具体来说，对于自学读者，在阅读教材时，应该边阅读、边实践。如果对于教材中的某些概念、语法存有疑问，应该立即编写程序验证。对于教材中的例题，应该先尝试进行编程，之后再看教材的答案。教材中出现的包、类，都应该通过 Java API 手册了解其主要的功能，并掌握主要方法的功能。所谓熟能生巧，一些 API 在看多了之后，自然而然就知道其功能了。在完全理解了主教材内容以后，再开始做实验和习题。

对于在校学习 Java 课程的学生，应按照教师讲课的进度，提前预习教材。所谓预习，并不是要完全看懂，如果都看懂了，就不必听课了。预习的目的是大致浏览一下新的内容，了解哪些是难点、重点，听课时就比较主动。上课之后要及时复习，然后再写作业。复习时要边看书边记笔记，这时一定要认真阅读书上的内容，并同时查看 Java API 手册，掌握书上用到的类及其方法的用法。教师可能不会在课堂上讲解书中的每一个例题，对于教师课上讲的例题和书上的例题，课后复习时都要阅读、上机实践，达到完全理解，并具备自己独立编写例题程序的能力。做到这些以后，再开始写作业。

当然，这只是针对大多数读者的一般建议，每个人还要根据自己的情况选择适合的方法。

二、本书的使用方法

每学习一章主教材内容，都应该及时通过实验和习题巩固知识，提高实践能力。本书中的实验，是针对主教材每一章的重点内容设计的最基本的实践任务，有详细的实验指导，很容易入手，应该首先完成。完成实验之后，可以根据自己的时间和教师的要求，选择部分或全部习题来做。

本书给出了全部习题的答案，这是为了方便没有教师指导的自学读者。但是不少读者在没有深入思考之前就急于看答案，这是有害无益的，这样做不仅不能真正提高自己的编程能力，还会扼杀自己的创造性思维能力。特别是对于答案代码较多的习题以及从第 6 章开始的习题，具有较好的应用性，应该自己先进行编程，这样才能真正掌握所学知识。自学读者纷纷来信希望给出习题解答，而大多数教师(包括我自己)都不希望学生看到习题解答。这个矛盾困扰了我很久，始终没有找到两全的解决方案，所以只好在这里给予建议。

当然，有些章的习题较多，如果读者没有时间全部做完，也可以将一部分习题解答作为例题来学习。值得强调的是，当看到自己不熟悉的类或者方法时，还是要查阅 API 手册，了

解其功能,并能在之后的编程中使用。

三、本套教材的思路与使用要点

本套教材是针对已经先修过 C/C++ 语言的读者。前 4 章基本上是讲解 Java 语言的语法,以及面向对象编程思想;从第 5 章开始,讲解了一些常用的专题。在已经掌握 C/C++ 语言的基础上,学习前 4 章的基本语法应该很容易。在讲解语法和面向对象编程思想的同时,例题中使用了很多 Java API,初学的读者对此会感到有些困难。我们的建议是:前 4 章的学习重点是语法和面向对象编程思想,并不是 API,能够看懂例题中的用法就行了,对于类库中更多的 API 可以先不去关注;从第 5 章开始,各专题中用到大量 Java API,读者需要学会自己查阅 API 文档,这也是编写 Java 程序需要具备的基本能力。

# 目 录

# 第 1 章

# Java 语言基础知识

## 要点导读

本章内容需要与配套的主教材《Java 语言程序设计》(第 3 版)第 1 章配合学习。

主教材第 1 章介绍了 Java 语言的基本知识。Java 是一种面向对象的语言。Java 语言的主要特点包括简单高效、面向对象、安全健壮、支持分布式、可移植、内置多线程支持、自动内存管理等特点。Java 提供了强大的类库支持,程序员可以使用丰富的 Java API 完成需要的功能。因此,程序员应该在编程过程中熟悉对 API 的使用,熟练掌握常用类和常用方法。

开发 Java 程序时,基本的开发工具是 JDK。JDK 中,常用的工具包括 javac、java、jdb 等。Java 源文件编译成字节码文件后,由运行在特定平台上的 Java 虚拟机解释并执行。Java 虚拟机的多平台性保证了 Java 字节码文件运行时的平台无关性。

本书采用 Java 13 作为文字内容和代码示例的基准版本,并使用集成开发环境 IntelliJ IDEA。

Java 提供了多种基本数据类型,并支持算术、逻辑、赋值、位运算等多种表达式。Java 提供隐式和显式类型转换机制,可将表达式转换为另外的类型。Java 中有较多的关键字,在后续的学习过程中将会相继介绍。

数组是由同类型的数据元素构成的一种数据结构,在编写 Java 程序时会经常使用。多维数组可以理解为数组的数组。

Java 中的流程控制结构主要有顺序结构、选择结构及循环结构三种。顺序结构即按照从上到下的顺序执行语句,没有转移及重复。选择结构是根据给定的条件成立与否,执行不同的语句或语句组。Java 的选择结构主要有 if 选择结构及 switch 选择结构两种。值得注意的是,使用 switch 选择结构时,不要漏写 break 语句。循环控制结构是在一定的条件下,反复执行某段程序的流程结构,被反复执行的程序称为循环体。Java 中提供的循环语句共有三种: for 语句、while 语句、do-while 语句。

## 实验 1　Java 简单程序设计

### 一、实验目的

(1) 学会编写简单的 Java 程序。

(2) 学会使用命令行方式编译运行 Java 程序。

（3）学会使用 IntelliJ 编译运行 Java 程序。

（4）学会数组的创建和使用。

（5）掌握三种流程控制语法，并熟练应用。

## 二、实验任务

### 1. 编写简单程序

编写简单的 Java 应用程序，输出“Hello World!”，并分别用命令行方式和 IntelliJ IDEA 编译运行。

### 2. 定义和使用数组

在 Java 程序中创建包含整数 1～10 的数组，并使用循环结构输出所有偶数。

### 3. 复习、熟悉流程控制的语法

写出下面程序段的执行结果。

```
public static void main(String[] args) {
    int result = 0;
    int j = -100;
    for (int i = 0; i < 100; i++) {
        if (i %3 == 0) {
            j = i;
        } else {
            j = i * 2;
            while (j > 0) {
                j--;
                result += j;
            }
        }
        if (i %3 == 1) {
            result--;
            continue;
        }
        result += i;
        if (result > 0) {
            break;
        }
    }
    System.out.println(result);
}
```

## 三、实验步骤

### 1. 用命令行编译运行 Java 应用程序

（1）在文本编辑工具（例如记事本）中新建一个文件，保存为“D:\Java\exp1\problem1\HelloWorld.java”。编辑文件 HelloWorld.java 的内容为：

```
public class HelloWorld {
    public static void main (String[] args) {
        System.out.println("Hello World!");
    }
}
```

（2）打开 Windows 中的命令行窗口，使用“d:”“cd *”等命令进入目录“D:\Java\exp1\problem1”，使用“ls”命令查看该目录下的文件。

（3）使用“javac HelloWorld.java”命令，编译已经写好的 Java 文件。使用“ls”命令查看在当前目录下是否生成了文件“HelloWorld.class”。

（4）使用“java HelloWorld”命令，执行已经生成的“HelloWorld.class”文件，查看输出内容。

**2. 在 IntelliJ IDEA 中创建新的 Java 项目**

（1）在 IntelliJ 中单击 Open 按钮选择已有工程目录“D:\Java\exp1\problem1”并打开；或单击 New Project 按钮新建工程（图 1-1），选择 Empty Project 选项，命名为“exp1”（作为本章所有实验的父目录），右键单击工程“exp1”，在菜单中依次选择 New→Directory 命令创建子目录并命名为“problem1”作为本题的工程目录（图 1-2）。然后新建 Java 类（图 1-3），命名为“HelloWorld”。按要求编写 HelloWorld.java 文件的代码（图 1-4）。

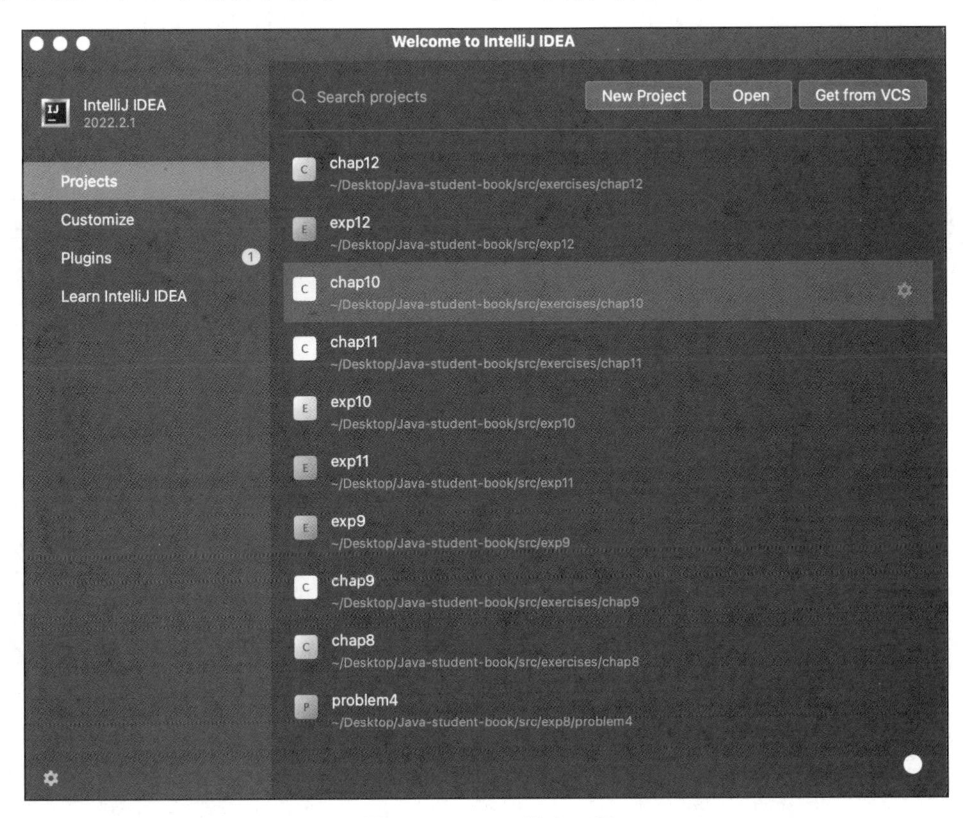


图 1-1　IntelliJ 新建工程

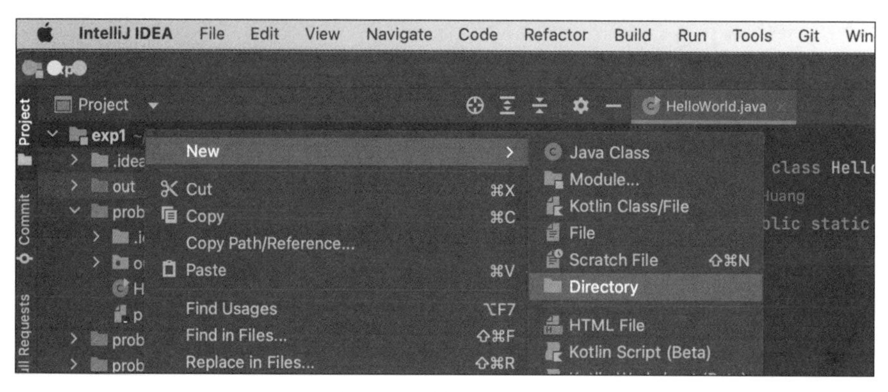


图 1-2　新建子目录

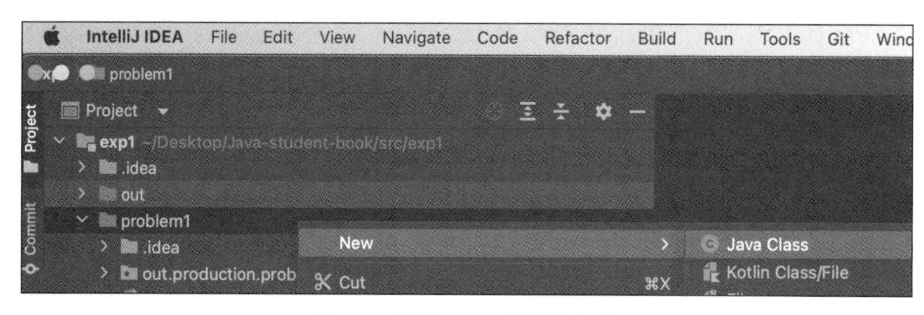


图 1-3　新建 Java 类代码文件

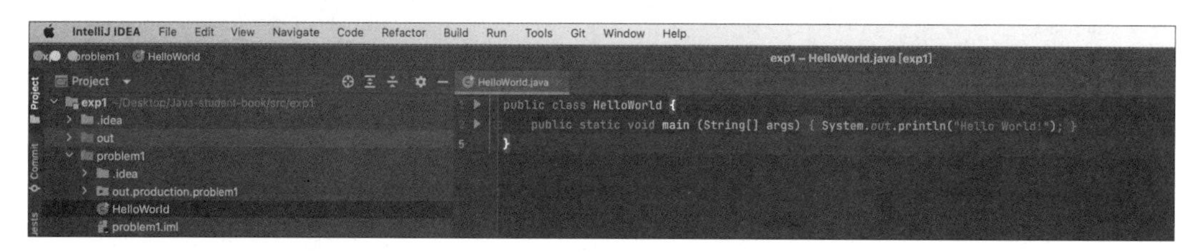


图 1-4　编写 HelloWorld 类代码

(2) 单击 Build 菜单下的 Build Project 命令(图 1-5),等待提示“Build completed successfully”。

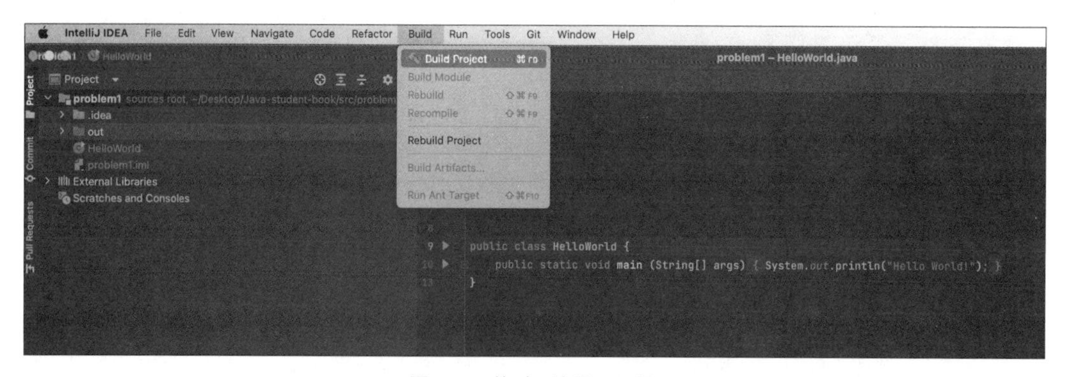


图 1-5　构建(编译)工程

（3）单击Run菜单下的Run 'HelloWorld.java'命令，查看输出结果（图1-6）。

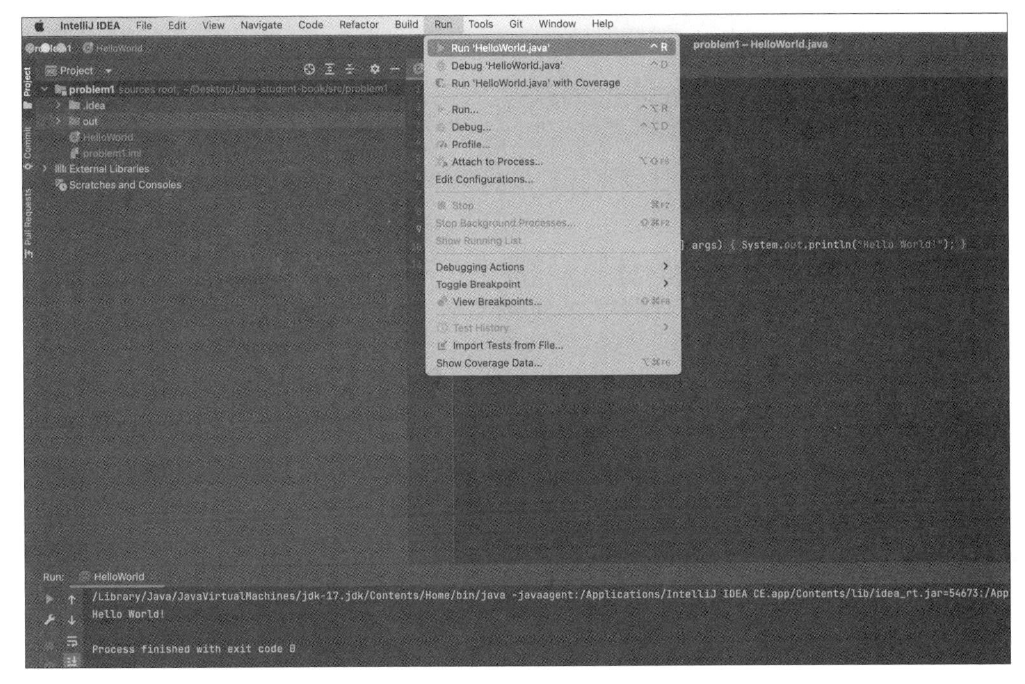

图1-6　运行

3. 数组的创建和使用

（1）同上题步骤，在项目“exp1”下新建子目录“problem2”，新建Java类命名为“MyArray”，在MyArray.java中编写主函数main()。

```
public class MyClass {
public static void main(String arg[]) {
...
    }
}
```

（2）在main()函数中初始化一个int型数组，包含1～10共10个元素。

```
int a[] = {1, 2, 3, 4, 5, 6, 7, 8, 9, 10};
```

（3）用for循环遍历数组（for (int i: a)），输出数组中所有偶数（可以用求余运算i%2==0来判断一个元素是否为偶数）。

4. 复习控制流程语法

（1）习惯使用大括号标明程序段。养成良好的编程习惯，即使分支、循环中只有一行代码，也最好用大括号括起来。

（2）按照缩进格式书写的代码更能清晰地看出程序流程的结构。

（3）目测实验任务3中的代码片断，计算出程序输出。

（4）目测完毕实验任务3后，可以编译、运行，以验证自己的计算。

## 习题解答

1.简述 Java 语言的发展历史。

解：

1991 年，Sun 公司的团队为了研究计算机在电子消费领域的应用，开发了一种新的语言，起名为 Java。

1995 年，Sun 公司发布了 Java 语言并开源。

1998 年，Sun 公司发布了 JDK 1.2 版本，将 Java 分成 J2SE、J2EE 和 J2ME 三个版本，面向不同的场景。

2004 年，Sun 公司发布了 JDK 1.5，后改名为 Java SE 5.0，增加了很多语言特性，还发布了新的企业级平台开发规范。

2006 年，Java SE 6 发布。

2009 年，Oracle 公司收购了 Sun 公司，但 Google 公司的 Android 系统极大地推动了 Java 的发展。

2011 年和 2014 年，Oracle 公司分别发布了 Java 7 和 Java 8，带来了 try-with-resource 语句、Lambda 表达式等新特性。

2017 年 9 月起，每半年发布一个 Java 新版本。

Oracle 公司在 2017 年宣布将 Java EE 移交给 Eclipse 基金会管理，Eclipse 于 2018 年将 Java EE 重命名为 Jakarta EE，而后发布了 Jakarta EE 8。

2. Java 语言主要有哪些特点？

解：

Java 语言的主要特点如下。

(1) 简单高效：Java 语言的语法和 C++ 语言很接近，而丢弃了 C++ 中指针、多继承等难以理解的内容，所以比较容易掌握。另外，Java 还提供了垃圾回收机制，使得内存无须程序员管理就能得到高效使用。

(2) 面向对象：Java 语言是一种优秀的面向对象的编程语言，提供了封装、继承和多态三大特性。

(3) 安全健壮：Java 语言通常用在网络环境中，因此提供了强大的安全机制以防止恶意代码的进攻。此外，Java 的异常处理机制、强类型机制和自动垃圾回收机制等为 Java 程序的健壮性提供了重要的保障。

(4) 分布式和可移植：Java 语言提供了网络应用编程接口以支持网络应用的开发，同时提供了 RMI(远程方法调用)机制以支持分布式应用开发。Java 程序是可移植的，因为 Java 程序在 Java 平台上被编译为平台无关的字节码格式，可以在任意实现了 Java 平台的系统中运行。

(5) 高性能和多线程：Java 的性能虽不如 C 等编译型语言，但在 JIT 编译器的帮助下，Java 方法在执行前被编译为本地机器代码，运行性能得到大幅提升。Java 语言支持多线程，并提供了线程间的同步和通信机制。

3. 参考主教材第 1 章的介绍搭建 Java 开发环境。

4. 修改 HelloJava.java 文件,使得控制台输出“Hello World!”。

解:

修改 HelloJava.java 文件的内容为:

```
public class HelloJava{
    public static void main(String[] args) {
        System.out.println("Hello World!");
    }
}
```

5. 设 $N$ 为自然数,$N!=1\times2\times3\times\cdots\times N$ 称为 $N$ 的阶乘,并且规定 0!=1。试编程计算 2!,4!,6! 和 10!,并将结果输出到屏幕上。

解:

新建 Exe1_5.java 文件,其内容为:

```
public class Exe1_5{
    public static void main(String[] args){
        int[] Ns = {2, 4, 6, 10};
        for(int N: Ns){
            int prod = 1;
            for(int i = 1; i <= N; i++){
                prod *= i;
            }
            System.out.println(prod);
        }
    }
}
```

程序的输出结果为:

```
2
24
720
3628800
```

6. 编写程序,接收用户从键盘上输入的三个整数 $x$、$y$、$z$,从中选出最大和最小者,并输出。

解:

新建 Exe1_6.java 文件,其内容为:

```
import java.util.Scanner;
public class Exe1_6{
    public static void main(String[] args){
        Scanner scanner = new Scanner(System.in);
        int maxVal = Integer.MIN_VALUE, minVal = Integer.MAX_VALUE;
        for(int i = 0; i < 3; i++){
            int x = scanner.nextInt();
```

```
            maxVal = Math.max(maxVal, x);
            minVal = Math.min(minVal, x);
        }
        System.out.println("Max: " + maxVal);
        System.out.println("Min: " + minVal);
    }
}
```

依次输入 1,2,3 时,输出结果为:

```
Max: 3
Min: 1
```

7. 求出 100 以内的素数,并将这些数在屏幕上 5 个一行地显示出来。

解:

新建 Exe1_7.java 文件,其内容为:

```
public class Exe1_7{
    public static void main(String[] args){
        final int N = 100;
        boolean[] isPrimes = new boolean[N+1];
        for(int i = 0; i <= N; i++)
            isPrimes[i] = true;
        for(int i = 2; i <= Math.sqrt(N); i++){
            if(!isPrimes[i])
                continue;
            for(int j = 2; ; j++){
                int num = i * j;
                if(num > N)
                    break;
                isPrimes[num] = false;
            }
        }
        int cnt = 0;
        for(int i = 2; i <= N; i++){
            if(isPrimes[i]){
                System.out.print("" + i + "");
                cnt++;
                if(cnt %5 == 0)
                    System.out.println();
            }
        }
    }
}
```

程序的输出结果为:

```
2  3  5  7 11
13 17 19 23 29
31 37 41 43 47
53 59 61 67 71
73 79 83 89 97
```

8. 使用 java.lang.Math 类，生成 100 个 0～99 的随机整数，找出它们之中的最大者及最小者，并统计大于 50 的整数个数。

解：

新建 Exe1_8.java 文件，其内容为：

```
public class Exe1_8{
    public static void main(String[] args){
        int minVal = 0, maxVal = 99;
        int minItem = maxVal + 1, maxItem = minVal - 1;
        int cnt = 0;
        for(int i = 0; i < 100; i++){
            int x = minVal + (int)(Math.random() * (maxVal - minVal + 1));
            //System.out.println(x);
            maxItem = Math.max(maxItem, x);
            minItem = Math.min(minItem, x);
            if(x > 50)
                cnt++;
        }
        System.out.println("Max item: " + maxItem);
        System.out.println("Min item: " + minItem);
        System.out.println("#int>50: " + cnt);
    }
}
```

程序的输出结果为：

```
Max item: 99
Min item: 0
#int>50: 48
```

9. 接收用户从键盘上输入的两个整数，求两个数的最大公约数和最小公倍数，并输出。

解：

新建 Exe1_9.java 文件，其内容为：

```
import java.util.Scanner;

public class Exe1_9{
    public static void main(String[] args){
        Scanner scanner = new Scanner(System.in);
        int x = scanner.nextInt(), y = scanner.nextInt();
        int a = Math.max(x, y), b = Math.min(x, y);//a > b
```

```
        //greatest common divisor
        int rmd = a %b;
        while(rmd != 0){
            a = b;
            b = rmd;
            rmd = a %b;
        }
        int gcd = b;

        //least common multiple
        int lcm = x * y / gcd;

        System.out.println("Greatest common divisor: " + gcd);
        System.out.println("Least common multiple: " + lcm);
    }
}
```

当输入 6 和 8 时,程序的输出结果如下。

```
Greatest common divisor: 2
Least common multiple: 24
```

# 第 2 章

## 类与对象的基本概念

### 要点导读

本章内容需要与配套的主教材《Java 语言程序设计》(第 3 版)第 2 章配合学习。

主教材第 2 章介绍了类与对象的基本概念。面向对象程序设计是一种围绕真实世界的概念来组织模型的程序设计方法,采用对象来描述真实世界中的某一个实体。对象在程序中是通过一种抽象数据类型来描述的,这种抽象数据类型称为类。类是对具有相同属性和功能的对象的抽象描述。可以认为,类描述了"一类"对象。

类的成员包括数据成员和方法成员,分别表示类的属性和行为。包用来对类进行组织。类通过 public 修饰符进行访问控制。声明为 public 的类可以被同一包以及不同包中的类访问,未声明为 public 的类只能被同一包中的类访问,不同包中的类不能互相访问。类成员通过 public、protected、private 修饰符进行访问控制。声明为 public、protected、无修饰符、private 的成员的可访问性逐步递减。

构造方法是类的方法成员中非常重要的成员,没有返回值,且方法名和类名相同。如果用户没有定义构造方法,系统会隐含生成一个默认构造方法,参数表为空。如果用户自定义了构造方法,就不会隐含生成默认构造方法。在生成一个对象时,系统会自动调用该类的构造方法为新生成的对象初始化。

枚举类型适合于描述一个有限的集合。

Java 还提供了注解的特性,注解通过元数据的形式为其所标注的程序提供一些额外的信息。Java 既提供特定的注解,也支持自定义注解。

### 实验 2　类与对象的基本概念

#### 一、实验目的

(1) 巩固类、方法、属性的基本概念。

(2) 学会编写简单的面向对象程序。

(3) 掌握类的 toString()方法的重载及使用。

#### 二、实验任务

编写一个分数的类 Fraction。此类包括两个整数型属性：分母和分子。定义该类构造方法及属性的存取方法。定义该类的化简方法，例如，4/8 化简为 1/2。定义该类的 toString 方法，输出化简后的结果(如果是整数，则只输出整数)。

为 Fraction 类编写两种共八个方法，实现加、减、乘、除运算：第一种方法的参数为一个 Fraction 对象，功能是当前 Fraction 对象与参数对象进行运算；第二种方法的参数为两个 Fraction 对象，返回这两个 Fraction 对象的运算结果。例如：

```
public void add(Fraction a) {
    //当前分数(即 this)加上 a,结果存储在 this 中
}
public Fraction add(Fraction a, Fraction b) {
    //返回 a 和 b 相加的结果
}
```

在一个单独的测试类 FractionTester 中，测试并验证这八个分数运算方法是否正确。

在 IntelliJ 和命令行中分别编译 Fraction 类和测试类，运行测试类，检查结果是否正确。

## 三、实验步骤

可以参考本章习题 11 解答中的复数类，完成分数类的代码编写。Fraction 类中的加减运算必要时需要使用通分操作，然后用两个操作数的分子和分母进行计算，计算完成后可以使用约分操作进行结果的化简。

### 1. 在命令行方式下编译、运行程序，验证结果

(1) 创建文件夹"problem1"作为包"problem1"的目录，在包"problem1"中创建 Java 文件 Fraction.java。

(2) 完成 Fraction 类的编写。

(3) 在包"problem1"下创建 Java 文件 FractionTester.java，在其 main()方法中编写测试代码。

注意：如果在某 Java 文件中使用 package 语句声明该文件位于某个包，此时使用命令行方式进入该 Java 文件所在目录，并使用 javac 命令编译该文件，会出现"NoClassDefFoundError"。例如，在本题中，由于 Fraction 类和 FractionTester 类都位于 problem1 包下，如果在 problem1 目录下编译 Fraction.java 文件，将会出现"NoClassDefFoundError"。正确的方法是应该在包所在的目录下编译该文件。

(1) 用如下命令分别编译 Fraction 和 FractionTester 两个类。

javac problem1/Fraction.java。

javac problem1/FractionTester.java。

(2) 用 java problem1.FractionTester 命令运行 FractionTester 的 main()方法。

### 2. 使用 IntelliJ 集成开发环境编写编译、运行程序，验证结果

(1) 在 IntelliJ 中单击 New Project 新建工程，选择 Empty Project 选项，命名为"problem1"。

(2) 选择 New→Java Class 命令依次新建两个 Java 类 Fraction 和 FractionTester，并完成代码的编写。

(3) 单击 Build→Build Project 命令构建,然后单击 Run→Run 'HelloWorld.java'命令运行。

## 习题解答

1. 什么是对象? 什么是类? 它们之间的关系如何?

解:

对象是包含现实世界物体特征的抽象实体,它反映了系统为之保存信息和(或)与它交互的能力。类是具有相同操作功能和相同的数据格式(属性)的对象的集合与抽象。类与对象的关系可以表述为:一个类是对一类对象的描述,是构造对象的模板,对象是类的具体实例。

2. 什么是面向对象的程序设计方法? 它有哪些基本特征?

解:

在面向对象的程序设计方法中,程序的基本组成单位是类。程序在运行时由类生成对象,对象之间通过发送消息进行通信,互相协作完成相应的功能。面向对象的程序设计方法的基本特征有:抽象、封装、继承、多态。

3. 在下面的应用中,找出可能用到的对象,对每一个对象,列出可能的状态及行为。

(1) 模拟航空预订系统交易的程序。

(2) 模拟银行交易的程序。

解:

(1) 对象:顾客;可能的状态及行为:姓名、性别、年龄、查找航班、预订机票。

对象:机票;可能的状态及行为:起飞时间、起点、终点、到达时间。

(2) 对象:顾客;可能的状态及行为:姓名、性别、身份证号、出生日期、查询余额、取款、存款。

对象:账户;可能的状态及行为:账号、开户时间、有效时间、取款、存款。

4. 解释类属性、实例属性以及它们的区别。

解:

类属性表示类中所有的对象都相同的属性,在声明时加上 static 修饰符。实例属性用来存储所有实例都需要的属性信息,不同实例的属性值可能会不同,在声明时不加 static 修饰符。二者的区别为:类属性为所有的对象拥有,实例属性为每个实例对象自己拥有。

5. 解释类方法、实例方法以及它们的区别。

解:

类方法表示类的功能,也称为静态方法,在方法声明时前面需加 static 修饰符,在使用时可以将类方法发送给类名,也可以发送给一个实例,建议采用前者。实例方法表示特定对象的行为,在声明时前面不加 static 修饰符,在使用时需要将实例方法发送给一个实例。二者的区别为:类方法在使用时既可以将类方法发送给类名,也可以发送给一个实例,实例方法在使用时需要将实例方法发送给一个实例。

6. 类的访问控制符有哪几种? 具体含义是什么?

解:

类的访问控制符有 public(公共类)及无修饰符(默认类)两种。当使用 public 控制符时，表示所有其他的类都可以使用此类；当没有修饰符时，则只有与此类处于同一包中的其他类可以使用此类。

7. 类成员的访问控制符有哪几种？它们对类成员分别有哪些访问限制作用？

解：

类成员的访问控制符有 public、private、protected 及无修饰符。用 public 修饰的成员表示是公有的，也就是它可以被其他任何对象访问（前提是对类成员所在的类有访问权限）。用 private 修饰的成员只能被这个类本身访问，在类外不可见。用 protected 修饰的成分是受保护的，只可以被同一类及其子类的实例对象访问。无修饰符时，表示相应的成员可以被所在包中的各类访问。

8. 简述构造方法的特点。

解：

构造方法（Constructor）是一种特殊的方法。构造方法具有和类名相同的名称，而且无返回值类型。系统在产生对象时会自动执行。构造方法主要有以下特点。

- 构造方法的方法名与类名相同。
- 构造方法没有返回类型（修饰符 void 也不能有）。
- 构造方法一般被声明为公有的（public）。
- 构造方法可以有任意多个参数。
- 构造方法的主要作用是完成对象的初始化工作。
- 构造方法不能在程序中显式地调用。
- 在生成一个对象时，系统会自动调用该类的默认构造方法为新生成的对象初始化。

9. 如果在类声明中声明了构造方法，系统是否还提供默认的构造方法？

解：

不提供。

10. 声明 Patient 类表示在门诊室中的病人，此类对象应包括：name（String 类型）、sex（char 类型）、age（int 类型）、weight（float 类型）、allergies（boolean 类型）等属性，以及属性对应的 getter()/setter()方法。在一个单独的类中，声明测试方法，并生成两个 Patient 对象，设置其状态并将其信息显示在屏幕上。下面是测试一个 Patient 的例子。

```
Patient april = new Patient();
april.setName("ZhangLi");
april.setSex('f');
april.setAge(33);
april.setWeight(154.72f);
april.setAllergies(true);
System.out.println("Name:   "+ april.getName());
System.out.println("Sex:     " + april.getSex());
System.out.println("Age:     " + april.getAge());
System.out.println("Weight: "+ april.getWeight());
System.out.println("Allergies:  "+ april.getAllergies());
```

声明并测试 toString()方法显示一个病人的 age、sex、name 及 allergies 属性。

解：

新建 Exe2_10.java 文件，其内容为：

```
class Patient {
    private String name;
    private char sex;
    private int age;
    private float weight;
    private boolean allergies;
    //构造方法
    public Patient(String name,char sex,int age,float weight,boolean allergies) {
        this.name=name;
        this.sex=sex;
        this.age=age;
        this.weight=weight;
        this.allergies=allergies;
    }
    //不带参数的构造方法
    public Patient() {
        this("",' ',0,0f,false);
    }
    //获得属性的方法
    public String getName() {
        return name;
    }
    public char getSex() {
        return sex;
    }
    public int getAge() {
        return age;
    }
    public float getWeight() {
        return weight;
    }
    public boolean getAllergies() {
        return allergies;
    }
    //修改属性的方法
    public void setName(String name) {
        this.name=name;
    }
    public void setSex(char sex) {
        this.sex=sex;
    }
    public void setAge(int age) {
        this.age=age;
    }
```

```
    public void setWeight(float weight) {
        this.weight=weight;
    }
    public void setAllergies(boolean allergies) {
        this.allergies=allergies;
    }
    //toString()方法,显示一个病人的各种属性
    public String toString() {
        String s = "病人" + name+"的属性如下:\n";
        s = s + "姓名:" + name + "\n";
        s = s + "性别:" + (sex=='f'?"女":"男") + "\n";
        s = s + "年龄:" + age + "\n";
        s = s + "体重:" + weight + "\n";
        s = s + "是否过敏:" + (allergies==true?"是":"不") + "\n";
        return s;
    }
}
//测试类
public class Exe2_10{
    public static void main(String[] args) {
        Patient april = new Patient();
        april.setName("ZhangLi");
        april.setSex('f');
        april.setAge(33);
        april.setWeight(154.72f);
        april.setAllergies(true);
        System.out.println("Name:    " + april.getName());
        System.out.println("Sex:     " + april.getSex());
        System.out.println("Age:     " + april.getAge());
        System.out.println("Weight: " + april.getWeight());
        System.out.println("Allergies:  " + april.getAllergies());
        Patient liu = new Patient("Liu Wu", 'm', 23, 135f, false);
        System.out.println(liu);
    }
}
```

程序输出结果为:

```
Name:   ZhangLi
Sex:    f
Age:    33
Weight: 154.72
Allergies:  true
病人 Liu Wu 的属性如下:
姓名: Liu Wu
性别: 男
```

```
年龄: 23
体重: 135.0
是否过敏: 不
```

11. 声明并测试一个复数类,其方法包括 toString()及复数的加、减、乘运算。

解:

新建 Exe2_11.java 文件,其内容为:

```
class ComplexNumber {
    private double real;
    private double image;
    //构造方法
    public ComplexNumber(double real,double image) {
        this.real=real;
        this.image=image;
    }
    //不带参数的构造方法
    public ComplexNumber() {
        this(0,0);
    }
    //修改属性的方法
    public void setReal(double real) {
        this.real=real;
    }
    public void setImage(double image) {
        this.image=image;
    }
    //读取属性的方法
    public double getReal() {
        return real;
    }
    public double getImage() {
        return image;
    }
    public String toString() {
        if (image == 0)
            return real + "";
        else if(image > 0)
            return real + "+" + image + "i";
        else
            return (real + "-" + Math.abs(image) + "i");
    }
    //加、减、乘三运算的定义,返回一个新的复数,当前复数不改变
    public static ComplexNumber add(ComplexNumber a, ComplexNumber b) {
        double r = a.getReal() + b.getReal();
        double i = a.getImage() + b.getImage();
        return new ComplexNumber(r, i);
    }
```

```
    public static ComplexNumber substract(ComplexNumber a, ComplexNumber b) {
        double r = a.getReal() - b.getReal();
        double i = a.getImage() - b.getImage();
        return new ComplexNumber(r, i);
    }
    public static ComplexNumber multiply(ComplexNumber a, ComplexNumber b) {
        double r = a.getReal() * b.getReal() - a.getImage() * b.getImage();
        double i = a.getReal() * b.getImage() + a.getImage() * b.getReal();
        return new ComplexNumber(r, i);
    }

    //加、减、乘三运算的定义,对当前对象执行加、减、乘操作
    public void add(ComplexNumber a) {
        real = real + a.getReal();
        image = image + a.getImage();
    }
    public void substract(ComplexNumber a) {
        real -= a.getReal();
        image -= a.getImage();
    }
    public void multiply(ComplexNumber a) {
        double r = real * a.getReal() - image * a.getImage();
        double i = real * a.getImage() + image * a.getReal();
        real = r;
        image = I;
    }
}
public class Exe2_11 {
    public static void main (String[] args) {
        ComplexNumber a = new ComplexNumber(1, 2);
        ComplexNumber b = new ComplexNumber(3, 4);
        ComplexNumber c = ComplexNumber.add(a, b);
        System.out.println("a + b = " + c);
        ComplexNumber d = ComplexNumber.substract(a, b);
        System.out.println("a - b = " + d);
        ComplexNumber e = ComplexNumber.multiply(a, b);
        System.out.println("a * b = " + e);
        c.add(a);
        System.out.println("After c.add(a), c = " + c);
        c.substract(a);
        System.out.println("After c.substract(a), c = " + c);
        c.multiply(a);
        System.out.println("After c.multiply(a), c = " + c);
    }
}
```

程序输出结果为：

```
a + b = 4.0+6.0i
a - b = -2.0-2.0i
a * b = -5.0+10.0i
After c.add(a), c = 5.0+8.0i
After c.substract(a), c = 4.0+6.0i
After c.multiply(a), c = -8.0+14.0i
```

# 第 3 章

# 类的重用

## 要点导读

本章内容需要与配套的主教材《Java 语言程序设计》(第 3 版)第 3 章配合学习。

主教材第 3 章介绍了类的重用。继承是面向对象程序设计的基石之一,是一种由已有类创建新类的机制。Java 要求声明的每个类都有超类,当没有显式指定超类时,超类隐含为 java.lang.Object 类。Java 不支持类的多重继承,只支持类的单继承,即每个子类只能有一个直接超类。一个对象继承的内容取决于此对象所属的类在类层次中的位置。一个对象从其所有的直接和间接超类中继承属性及行为。

隐藏和覆盖是指子类对从超类继承来的属性变量及方法可以重新定义。如果子类对从超类继承来的属性变量重新定义,则从超类继承的属性将被隐藏。如果子类在声明方法时,使用相同的方法名及参数表,但执行不同的功能,这种情况称为方法覆盖。

有继承时的构造方法遵循以下的原则:子类不能从超类继承构造方法;好的程序设计方法是在子类的构造方法中调用某一个超类构造方法;super 关键字也可以用于构造方法中,其功能为调用超类的构造方法;如果在子类的构造方法的声明中没有明确调用超类的构造方法,则系统在执行子类的构造方法时会自动调用超类的默认构造方法(即无参的构造方法);如果在子类的构造方法的声明中调用超类的构造方法,则调用语句必须出现在子类构造方法的第一行。

Object 类是 Java 程序中所有类的直接或间接超类,也是类库中所有类的超类,处在类层次最高点。如果两个对象具有相同的类型及相同的属性值,则称这两个对象相等(equal);如果两个引用变量指向的是同一个对象,则称这两个对象同一(identical)。

final 类是指被 final 修饰的类。final 类不可能有子类。final()方法是被 final 修饰的方法,不能被当前类的子类覆盖。

抽象类是指被 abstract 修饰的类,表示不能使用 new 进行实例化,即不能有具体实例对象。抽象类既可以包含抽象方法,也可以包含非抽象方法。抽象方法使用 abstract 修饰,有方法的声明,而没有方法的实现,并将在抽象类的子类中实现。不能在非抽象的类中声明抽象方法。

泛型的本质是参数化类型,即所操作的数据类型被指定为一个参数。这种数据类型的指定可以使用在类、接口以及方法中,分别称为泛型类、泛型接口和泛型方法。通配符泛型用参数“?”表示,代表任意一种类型。有限制的泛型是指对类型参数加上一定的限制,例如,必须是某类的子类或者实现了某接口。

Java 的类中可以有其他类的对象作为成员,这便是类的组合。

Java 提供了用于语言开发的类库,称为应用程序编程接口(Application Programming Interface,API),分别放在不同的包中。Java 提供的包主要有:java.lang、java.io、java.math、java.util、java.applet、java.awt、java.awt.datatransfer、java.awt.event、java.awt.image、java.beans、java.net、java.rmi、java.security、java.sql 等。

# 实验 3　类的重用

## 一、实验目的

(1) 理解类的组合与继承,知道何时使用哪种重用机制。

(2) 了解 final 类、final()方法、抽象类、抽象方法的概念。

(3) 熟练掌握主教材第 3 章提到的 Java 基础类库中的一些常见类,学会查阅 JavaDoc。

(4) 了解 Java 包的概念。知道为什么要用包,养成良好的命名习惯。

(5) 初步了解 JAR 文件的概念和 jar 命令的格式。

## 二、实验任务

### 1. 继承关系举例

举出两个例子,各包含一个超类和它的三个子类。对于超类,要给出两个成员变量,一个方法的定义。要求两个例子一个是具体事物,另一个是抽象事物。不要使用教材中已有的例子,最好能和自己所学的专业有些关系。

例如:

具体事物:主教材 3.4.1 节中的 Shape、Circle、Triangle、Rectangle 的例子。超类成员变量可以有周长、面积。方法可以有 draw(),功能是把图形画出来。

抽象事物:超类,如银行相关交易类。子类,如存款交易、贷款交易、外汇交易。超类成员变量可以有交易时间、交易金额。方法可以有 save(),功能是把交易记录保存。

### 2. equals()方法

下面的代码实现了复数类和复数的减法。equals()方法是比较两个对象是否相等,每个 Java 类都有一个继承自 Object 类的默认的 equals()方法。下面的类有 main()方法,那它的执行结果是什么?为什么会有这样的结果?这个执行结果真的是我们期待的结果吗?重写 ComplexNumber 类的 equals()方法,使得它真正能够比较两个复数是否相等。你添加的 equals()方法应该使这个 main()方法返回应有的结果(true true true true false false)。

注意:equals()方法的参数是 Object 对象,所以应该考虑如果输入其他对象(例如 String)或者空对象(null)应该如何处理,应该避免 equals()方法出现什么样的异常。也可以使用自己正确编写的复数类。

```
public class ComplexNumber {
    double real = 0;

    double imagine = 0;
```

```
    public ComplexNumber(double real, double imagine) {
        this.real = real;
        this.imagine = imagine;
    }

    public ComplexNumber minus(ComplexNumber operand) {
        return new ComplexNumber(this.real - operand.real, this.imagine
                - operand.imagine);
    }

    public static void main(String[] args) {
        ComplexNumber complex1 = new ComplexNumber(2.02d, 3.1d);
        ComplexNumber complex2 = new ComplexNumber(2d, 3d);
        ComplexNumber complex3 = complex2;
        ComplexNumber complex4 = new ComplexNumber(2d, 3d);
        ComplexNumber complex5 = new ComplexNumber(0.02d, 0.1d);
        ComplexNumber complex6 = complex1.minus(complex2);

        System.out.println(complex2 == complex3);
        System.out.println(complex2.equals(complex3));
        System.out.println(complex2.equals(complex4));
        System.out.println(complex6.equals(complex5));
        System.out.println(complex1.equals(null));
        System.out.println(complex1.equals(new String("abc")));
    }
}
```

### 3. Java 类库常用包使用练习 1

输入一个包括 4 个小数的字符串，数之间用分号分隔，格式为“a;b;c;d”。计算如下表达式：

$$\sin a \times \cos b \times \sqrt{c^d}$$

返回与结果最接近的整数并按照格式输出计算时间。例如：

```
输入:
0.5;-0.8;3;6.3
输出:
2021-10-28 8:00:06
result = 11
```

如果输入的数字格式不正确，输出“Invalid Input”字符串。

提示：使用 java.util.StringTokenizer，java.text.SimpleDateFormat，java.lang.Math 类。

### 4. 定义并实现一个记账软件的类

设计、定义、实现一个交易管理软件中会使用到的类。该软件可以记录和管理如下信息。

(1) 交易,包括日常收支交易、转账交易、投资买卖交易(外汇、股票)。

(2) 人员机构,包括人员、机构。

(3) 账户,包括活期账户、定期账户、信用卡账户、投资买卖交易账户。

定义所需类的超类、子类;合理地使用抽象类、final 类;合理地定义每个类的成员变量及其类型;合理地给出几个类的方法(无须实现方法)。

**5. Java 类库常用包使用练习 2**

用两种方法实现 encode()、decode()方法给字符串加解密。输出原始字符串、加密后的字符串和解密后的字符串。测试字符串"We will break out of prison at dawn"。

方法一:按照下列规则对字符串进行替换加密,包括大小写字母,"_"代表空格字符。

a b c d e f g h i j k l m n o p q r s t u v w x y z _

v e k n o h z f _ i l j x d m y g b r c s w q u p t a

方法二:就像在 JavaDoc 中看到的一样,计算机生成的是伪随机数。给 Random 对象设置一个固定的种子 Random.setSeed(long),它就能输出一个固定的随机序列出来。控制这个随机序列的范围,然后用这个序列和字符串序列进行可逆的运算,求得加密后的串。解密时通过相同的 seed 再现那个随机序列,并使用逆运算恢复原来的字符串。

提示:字符串里的字符是以数字保存的,它们可以进行加减求余等的运算。

读者也可以使用自己的方法来替代方法二并附加简要说明。可以考虑如何将这个功能打包成一个 jar 文件给别人使用。

## 三、实验步骤

**1. 复习理解类的继承并创建本章的工程**

(1) 复习、理解类的继承,区分类的继承与组合关系。

(2) 描述出实验任务 1 要求的例子。

(3) 使用 IntelliJ IDEA 创建工程 exp3,本章的包都在 exp3 中。

说明:此后各章的实验代码都分别以"exp 章号"为工程名,创建工程的步骤在后续各章的实验步骤叙述中将略去。后续各章的实验不再使用命令行编译的方式,建议使用 IntelliJ IDEA 作为开发环境,读者也可以选择自己熟悉的集成开发环境。

**2. 实现 Complex 类的 equals()方法**

(1) 创建包"problem2",在"problem2"包下创建 ComplexNumber.java,实现实验任务 2 中的代码。

(2) 理解 Object 类在整个 Java 语言中的地位,以及它所提供的常用方法。

(3) 不实现 equals()方法,使用 Object 的默认实现,运行 main()方法,看所得结果与自己的预测是否一致。

(4) 按照要求给 ComplexNumber 类实现 equals()方法,并通过 main()方法的测试,返回正确结果。或者使用自己编写的 ComplexNumber 类。

注意:浮点数在计算机中的表示是不精确的。因此,如果要判断两个浮点数是否相等,最好不要使用符号"==",而是要判断二者的差的绝对值是否很小,例如,是否小于 $10^{-5}$,如果很小,则可以认为这两个浮点数是相等的。

3. 熟悉常用 Java 类库一

(1) 打开 Java API 文档,熟悉各部分的内容,能够快速找到所需的类。

(2) 查看 java.util.StringTokenizer, java.text.DateFormat, java.lang.Math 类,熟悉其方法。

(3) 建立名为“problem3”的包,在包 problem3 下创建 Compute.java,实现实验任务 3 的要求。

4. 设计并实现记账软件的类

(1) 按照实验任务 4 的要求,设计记账软件的类。

(2) 考虑设计的合理性,合理使用抽象类、final 类等。

(3) 设计类的成员变量、常用方法。

(4) 新建 Java 包 MyAccount,实现自己设计的类。

记账软件的参考类图见图 3-1。

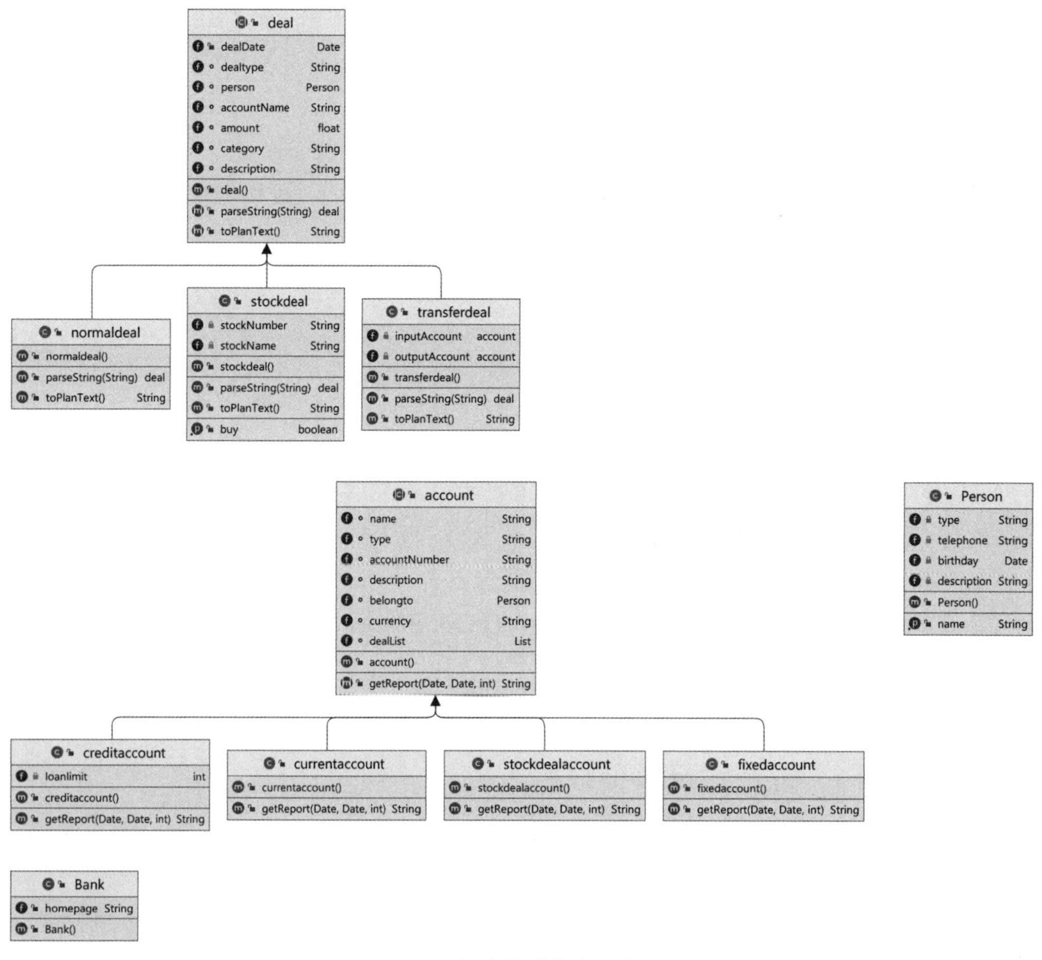


图 3-1　记账软件的参考类图

5. 熟悉常用 Java 类库二

(1) 打开 Java API 文档,查看 java.lang.StringBuffer、java.util.HashMap 等类,熟悉其

方法。也可以使用其他类来协助完成实验要求。

(2) 新建名为"problem5"的包,在 problem5 包下创建 StringEncryptor.java。创建方法 encode1()、decode1()、encode2()、decode2(),实现实验任务要求。自己在 main()里写测试方法,以验证正确性。

提示:检查加密后的字符串能否通过解密复原。

(3) 复习 jar 文件的概念。可以尝试写一个处理 String args[]参数的 main()方法。然后将这个类打包成 jar 包,供他人使用。

## 习题解答

1. 子类将继承超类所有的属性和方法吗?为什么?

解:

子类不从超类继承构造方法,但可通过"super.方法名"访问;除此之外,子类从超类继承其他所有属性和方法,但 private 属性/方法不能直接访问。

2. 方法覆盖与方法重载有何不同?

解:

方法覆盖是指如果子类不需要使用从超类继承来的方法的功能,则可以声明自己的方法。在声明时,使用相同的方法名及参数表,但执行不同的功能。而重载是指名字一样但参数表不一样的方法。二者的不同主要在于:方法覆盖时,子类的参数表和父类一样,方法重载时,参数表不一样。

3. 泛型的本质是什么?泛型可以使用在哪些场合?

解:

其本质是参数化类型,即所操作的数据类型被指定为一个参数。

泛型可以使用在类、接口以及方法中,分别称为泛型类、泛型接口和泛型方法。

4. 声明两个带有无参构造方法的类 A 和 B,声明 A 的子类 C,并且声明 B 为 C 的一个成员,不声明 C 的构造方法。编写测试代码,生成类 C 的实例对象,并观察结果。

解:

新建 Exe3_4.java 文件,其内容为:

```
class A {
    public A() {
        System.out.println("In Constructor of A");
    }
}
class B {
    public B() {
        System.out.println("In Constructor of B");
    }
}
class C extends A {
    B b;
}
```

```
public class Exe3_4 {
    public static void main(String[] args) {
        C c = new C();
    }
}
```

程序的输出结果为：

```
In Constructor of A
```

从输出结果可以看出，在通过默认构造方法创建类 C 的对象 c 时，自动调用了类 C 的超类 A 的无参数构造方法，而没有初始化 C 的成员 B。

5. 声明一个超类 A，它只有一个非默认构造方法；声明 A 的子类 B，B 具有默认方法及非默认方法，并在 B 的构造方法中调用超类 A 的构造方法。

解：

新建 Exe3_5.java 文件，其内容为：

```
class A {
    int i;
    A (int i) {
        this.i = i;
        System.out.println("In A (int i), i = " + i);
    }
}
class B extends A {
    int j;
    B () {
        super(0);
        System.out.println("In B ()");
        j = 0;
    }
    B (int j) {
        super(j);
        System.out.println("In B (int j)");
        this.j = j;
    }
}
public class Exe3_5 {
    public static void main (String[] args) {
        B b1 = new B();
        System.out.println("********************");
        B b2 = new B(5);
    }
}
```

程序的输出结果为：

```
In A (int i), i = 0
In B ()
*********************
In A (int i), i = 5
In B (int j)
```

读者可以尝试在类 B 的无参数构造方法中不调用“super(0)”，此时会有编译错误，即找不到类 A 的无参数构造方法。

6. 声明一个类，它具有一个方法，此方法被重载三次，派生一个新类，并增加一个新的重载方法，编写测试类验证四个方法对于子类都有效。

解：

新建 Exe3_6.java 文件，其内容为：

```
class Base {
    int i;
    public void setValue(int i) {
        this.i = i;
        System.out.println("In setValue(int i), i = " + this.i);
    }
    public void setValue(float f) {
        this.i = (int)f;
        System.out.println("In setValue(float f), i = " + this.i);
    }
    public void setValue(double d) {
        this.i = (int)d;
        System.out.println("In setValue(double d), i = " + this.i);
    }
}
class Sub extends Base{
    //在派生类中重载 setValue()方法
    public void setValue(byte b) {
        this.i = (int)b;
        System.out.println("In setValue(byte b), i = " + this.i);
    }
}
public class Exe3_6 {
    public static void main(String args[]) {
        Sub s = new Sub();
        s.setValue(2);
        s.setValue(3.5f);
        s.setValue(2.8d);
        s.setValue((byte)8);
    }
}
```

程序的输出结果为：

```
In setValue(int i), i = 2
In setValue(float f), i = 3
In setValue(double d), i = 2
In setValue(byte b), i = 8
```

7. 声明一个具有 final 方法的类，声明一个子类，并试图对这个方法进行覆盖(override)，观察会有什么结果。

解：

出现编译错误，因为 final()方法不能在子类中对其进行覆盖。

8. 声明一个 final 类，并试图声明其子类，观察会有什么结果。

解：

将会出现编译错误，并提示无法从 final 类继承。

9. 什么是抽象类？抽象类中是否一定要包括抽象方法？

解：

抽象类就是不能使用 new 进行实例化的类，即没有具体实例对象的类。抽象类中不一定要包括抽象方法。

10. this 和 super 分别有哪些特殊含义？都有哪些用法？

解：

关键词 this 是当前对象的引用，关键词 super 表示超类。二者都可以用于：调用本类或超类的方法，访问本类或超类的属性，调用本类或超类的构造方法。

11. 完成下面超类及子类的声明：

(1) 声明 Student 类。

属性包括：学号、姓名、英语成绩、数学成绩、计算机成绩和总成绩。

方法包括：构造方法、getter()方法、setter()方法、toString()方法、equals()方法、compare()方法(比较两个学生的总成绩，结果可以是大于、小于、等于)，sum()方法(计算总成绩)和 testScore()方法(计算评测成绩)。

注：评测成绩可以取三门课成绩的平均分，另外，任何一门课的成绩的改变都需要对总成绩进行重新计算，因此，在每一个 setter()方法中应调用 sum()方法计算总成绩。

(2) 声明 StudentXW(学习委员)类为 Student 类的子类。

在 StudentXW 类中增加责任属性，并覆盖 testScore()方法(计算评测成绩，评测成绩=三门课的平均分+3)。

(3) 声明 StudentBZ(班长)类为 Student 类的子类。

在 StudentBZ 类中增加责任属性，并重写 testScore()方法(计算评测成绩，评测成绩=三门课的平均分+5)。

(4) 声明测试类，生成若干个 Student 类、StudentXW 类及 StudentBZ 类对象，并分别计算它们的评测成绩。

解：

新建 Exe3_11.java 文件，其内容为：

```
class Student {
    int ID;
    String name;
    int englishScore;
    int mathScore;
    int computerScore;
    int sumScore;
    //构造方法
    Student(int ID, String name, int englishScore, int mathScore,
            int computerScore) {
        this.ID = ID;
        this.name = name;
        this.englishScore = englishScore;
        this.mathScore = mathScore;
        this.computerScore = computerScore;
        sum();
    }
    //设置属性的方法
    public void setID(int ID) {
        this.ID = ID;
    }
    public void setName(String name) {
        this.name = name;
    }
    public void setEnglishScore(int englishScore) {
        this.englishScore = englishScore;
        sum();
    }
    public void setMathScore(int mathScore) {
        this.mathScore = mathScore;
        sum();
    }
    public void setComputerScore(int computerScore) {
        this.computerScore = computerScore;
        sum();
    }
    //得到属性的方法
    public int getID() {
        return ID;
    }
    public String getName() {
        return name;
    }
    public int getEnglishScore() {
```

```
        return englishScore;
    }
    public int getMathScore() {
        return mathScore;
    }
    public int getComputerScore() {
        return computerScore;
    }
    public String toString() {
        String s = name;
        s += "(学号为" + ID + ")的成绩如下: \n";
        s += "英语: " + englishScore + "\n";
        s += "数学: " + mathScore + "\n";
        s += "计算机: " + computerScore + "\n";
        s += "总分: " + sumScore + "\n";
        return s;
    }
    @Override
    public boolean equals(Object o) {
        Student s = (Student) o;
        //当二者的总成绩相等时,返回 true
        return this.sumScore == s.sumScore;
    }
    public int compare(Student s) {
        //大于时返回 1,等于时返回 0,小于时返回-1
        if (this.sumScore > s.sumScore)
            return 1;
        else if(this.sumScore == s.sumScore)
            return 0;
        else
            return -1;
    }
    public void sum() {      //计算总分
        sumScore = englishScore + mathScore + computerScore;
    }
    public int testScore() {
        return sumScore / 3;
    }
}
class StudentXW extends Student{
    int duty;
    StudentXW(int ID, String name, int englishScore, int mathScore,
            int computerScore, int duty) {
        super(ID, name, englishScore, mathScore, computerScore);
        this.duty = duty;
```

```
    }
    @Override
    public int testScore() {
        return sumScore / 3 + 3;                //平均分加上 3
    }
}
class StudentBZ extends Student{
    int duty;
    StudentBZ(int ID, String name, int englishScore, int mathScore,
            int computerScore, int duty) {
        super(ID, name, englishScore, mathScore, computerScore);
        this.duty = duty;
    }
    @Override
    public int testScore() {
        return sumScore / 3 + 5;                //平均分加上 5
    }
}
//测试类
public class Exec3_11 {
    public static void main(String[] args) {
        Student s = new Student(20100023, "张三", 85, 92, 90);
        StudentXW xw = new StudentXW(20100015, "李四", 80, 90, 95);
        StudentBZ bz = new StudentBZ(20100005, "王五", 82, 85, 88);
        System.out.print(s);
        System.out.println("评测成绩: " + s.testScore());
        System.out.print(xw);
        System.out.println("评测成绩: " + s.testScore());
        System.out.print(bz);
        System.out.println("评测成绩: " + s.testScore());
    }
}
```

程序的输出结果为：

```
张三(学号为 20100023)的成绩如下:
英语: 85
数学: 92
计算机: 90
总分: 267
评测成绩: 89
李四(学号为 20100015)的成绩如下:
英语: 80
数学: 90
计算机: 95
```

```
总分: 265
评测成绩: 89
王五(学号为 20100005)的成绩如下:
英语: 82
数学: 85
计算机: 88
总分: 255
评测成绩: 89
```

12. 包有什么作用? 如何声明包和引用包中的类?

解:

包可以管理命名空间,可以解决同名类的冲突问题;也可以组织相关的类,并控制类的访问权限。

使用 package 关键字声明包。使用 import 关键字引用包中的类。

# 第 4 章

# 接口与多态

## 要点导读

本章内容需要与配套的主教材《Java 语言程序设计》(第 3 版)第 4 章配合学习。

主教材第 4 章介绍了接口与多态。接口可以看作一个“纯”抽象类,Java 中的接口是为了设计的多继承。接口中所有方法都是抽象的,这些抽象方法由实现这一接口的不同类来具体完成。利用接口构造类的过程,称为接口的实现,实现接口使用 implements 关键字。接口与一般类一样,均可以通过扩展来派生出新的接口。

类型转换也称为塑型。对象只能被塑型为:任何一个超类类型,或对象所属类实现的一个接口类型,或它自己所属的类。隐式类型转换是指自动进行类型转换。对于基本数据类型,相容类型之间存储容量低的自动向存储容量高的类型转换;对于引用变量,当一个类需要被塑型成更一般的类(超类)或接口时,系统会进行隐式类型转换。显式类型转换是强制在代码中将一个对象转换为另一个类型的对象。塑型主要应用于赋值转换、方法调用转换、算术表达式转换、字符串转换等场合。

将一个方法调用同一个方法主体连接到一起就称为绑定。根据绑定的时期不同,可将绑定分为“前期绑定”和“后期绑定”两种。在程序运行以前执行的绑定叫作前期绑定,在运行期间进行的绑定叫作后期绑定。后期绑定也叫作“动态绑定”或“运行期绑定”。

利用向上塑型技术,一个超类的引用变量可以指向不同的子类对象,而利用动态绑定技术,可以在运行时根据超类引用变量所指对象的实际类型执行相应的子类方法,从而实现多态性,即不同类型的对象可以响应相同的消息。

## 实验 4　接口与多态

### 一、实验目的

(1) 理解接口、类型转换、多态的概念并能熟练应用。

(2) 熟练掌握构造方法的调用顺序,理解需要注意的问题。

### 二、实验任务

#### 1. 继承时的多态

目测给出下面代码的执行输出结果。简单解释每行输出的原因。

```
class Car extends Vehicle {
    public Car() {
        System.out.println("A new Car.");
    }
    public void drive() {
        System.out.println("Car is driven");
    }
    public static void brake() {
        System.out.println("Car is braked");
    }
}

public class Vehicle {
    class Bus extends Vehicle {
        public Bus() {
            System.out.println("A new Bus.");
        }
        public void drive() {
            System.out.println("Bus is driven");
        }
    }

    public Vehicle() {
        System.out.println("A new Vehicle.");
    }

    public void drive() {
        System.out.println("Vehicle is driven");
    }

    public static void brake() {
        System.out.println("Vehicle is braked");
    }

    public void test() {
        Vehicle vc = new Car();
        Bus vb = new Bus();
        drive();
        vc.drive();
        vb.drive();
        vc.brake();
        vb.brake();
    }

    public static void main(String[] args) {
```

```
        Vehicle v = new Vehicle();
        v.test();
    }
}
```

**2. Comparable 接口和 Comparator 接口的使用**

给 myaccount.model.Deal 实现 java.lang.Comparable 接口，实现 Deal 类的默认比较功能，定义 Deal 的顺序为日期(dealDate)靠前的小，靠后的大。另外，实现依照金额(amount)大小给 Deal 排序的类：myaccount.util.DealAmountComparator，使其实现 java.util.Comparator 接口，这次定义 Deal 的顺序为金额大的靠前，小的靠后。在任务 3 中验证所实现的两个排序功能。

**3. 继续完善记账软件：超类的方法在不同子类中的不同实现**

在实验 3 的实验任务 4 基础上继续完善"记账软件"，给 myaccount.Deal 的三个子类实现它们的 toPlainText()方法。根据子类类型返回不同形式的字符串，包括类型、日期、数额等成员变量。例如，一个 NormalDeal 应该返回类似"＜Normal＞ 2006-11-01　￥220.5 description"的字符串。分别实现它们的含参数构造函数，使得这些子类能够根据不同的值来初始化。

编写一个测试程序，要求：

(1) 在程序中定义一个 Deal 的数组。

(2) 随机生成 10 个 Deal 子类放在数组中，要求随机生成种类、日期(dealDate)、数额(amount)。

(3) 遍历输出整个数组(使用 toPlainText()方法)。

(4) 使用默认的排序方案(日期排序)将数组排序。

(5) 遍历输出整个排序后的数组。

(6) 使用按金额排序方案将数组排序。

(7) 再次遍历输出整个排序后的数组。

请体会多态、类型转换、绑定等概念在本题中的应用。

本题无须实现排序的逻辑，使用上面任务 2 实现的接口即可。要对数组排序可以使用 java.util.Arrays 类的 Arrays.sort(Object[])和 Arrays.sort(Object[], Comparator)对数组进行排序。也可以使用 java.util.ArrayList 来代替数组，它也提供了相应的 sort()方法。

## 三、实验步骤

**1. 复习多态、绑定、构造方法等概念**

(1) 给出实验任务 1 代码的执行结果。

(2) 根据结果逐行做简要解释。写入 txt 文档。

(3) 题目中定义了几个有继承关系的类，还有静态和非静态的方法。复习关于构造方法与多态等相关内容，给出执行 main()方法后的输出结果，并解释。

(4) 在工程 exp4 中创建 Java 包，名称为"problem1"。在"problem1"包下创建 Vehicle.java，复制实验任务 1 代码到 Vehicle.java 中。运行程序，检查自己的理解是否正确。

**2. 在记账软件 MyAccount 项目中实现接口**

（1）将实验 3 完成的 MyAccount 包对应文件夹“MyAccount”复制到本章工程“exp4”目录下，在此基础上开始完成实验任务 2。

（2）实现 Comparable 接口。使 myaccount.model.Deal 类实现 java.lang.Comparable 接口，实现该接口定义的 compareTo()方法。根据名字理解，Comparable 就是可比较的。一个类如果实现了 Comparable 接口，它就是一个可被比较的类。思考一下为什么这个接口的方法要这样定义。

（3）实现 Comparator 接口。新建一个 myaccount.util.DealAmountComparator 类，使其实现 java.util.Comparator 接口，实现该接口定义的 compare()方法。顾名思义，Comparator 就是一个比较器，它可以用来比较某些类型的对象。思考一下为什么这个接口的方法要这样定义。

（4）对比 Comparable 和 Comparator 两个接口，思考它们有什么不同，结合实验任务 3，理解定义接口的目的和实际作用。

**3. 完成改进版的记账软件**

（1）打开 MyAccount 项目。

（2）按照题目给 Deal 的子类实现 toPlainText()方法。

（3）按照题目给 Deal 的子类实现带参数的构造方法。

（4）新建类 myaccount.app.TestDeal。在它的 main()函数中实现测试方法。

（5）观察输出是否符合预期。

改进的记账软件参考类图见图 4-1。

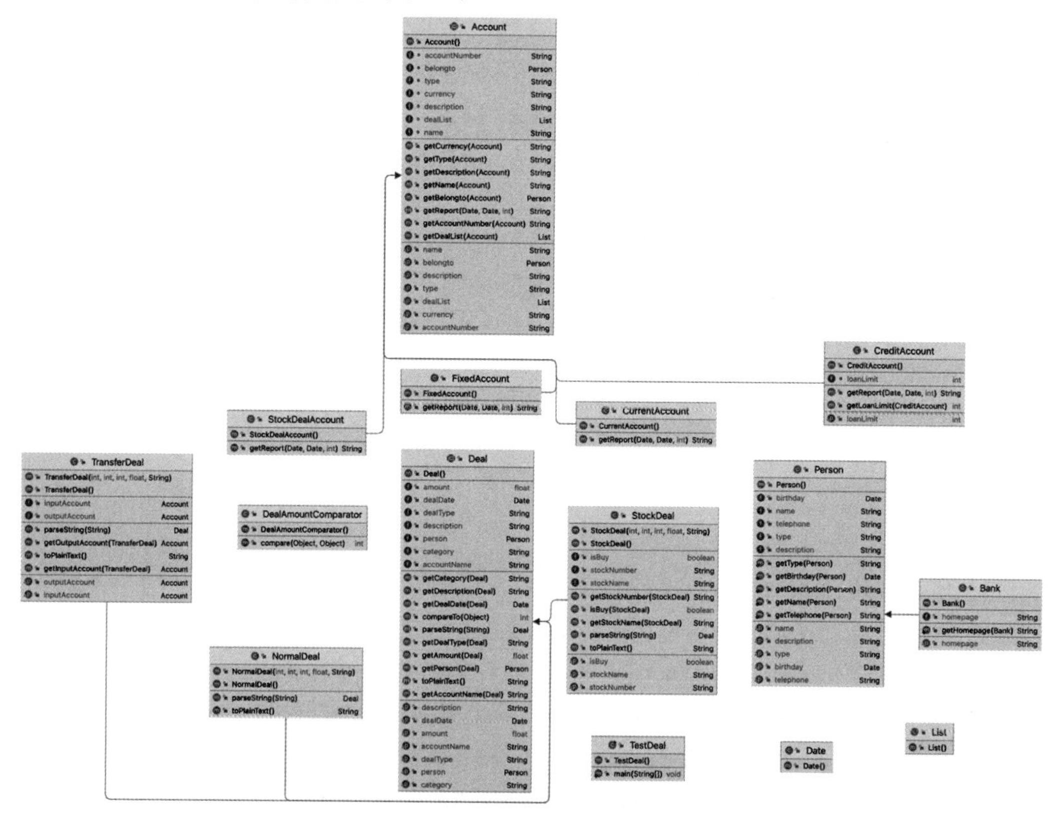


图 4-1　记账软件的参考类图

## 习题解答

1. 什么是接口？接口起什么作用？接口与抽象类有何区别？

解：

接口在语法上有些类似于抽象类，在其中声明了若干抽象方法和常量。其主要作用是帮助实现类设计的多重继承的功能。接口和抽象类的区别：接口中的所有方法都是抽象的，而抽象类中可以有非抽象的方法。

2. 编程证明接口中的属性都隐含为 static 及 final，所有的方法都为 public。

解：

建立包 shape，在其中新建 Shape2D.java 文件，其内容为：

```
package shape;
public interface Shape2D {
    double pi = 3.14;
    double area();
}
在另外的包中新建 Exe4_2.java 文件，其内容为:
import shape.Shape2D;
class Circle implements Shape2D{
    double r;
    Circle(double r) {
        this.r = r;
    }
    //double area() {return pi * r * r;}
    public double area() {
        return pi * r * r;
    }
}
public class Exe4_2{
    public static void main(String[] args) {
        Circle c = new Circle(2);
        Shape2D s = (Shape2D);
        System.out.println("s.area is" + s.area());
        //Exe4_2 和 Shape2D 属于不同包,且不是 Shape2D 的子类,但可以访问 s.area()
        //说明 Shape2D 中的 area()方法是 public 的
        System.out.println("c.area is " + c.area());
        System.out.println("Circle.pi = " + Circle.pi);
        //Circle.pi = 3.15; 不能为 Circle.pi 赋值
    }
}
```

Exe4_2.java 文件可以正常编译和运行，说明接口 Shape2D 中的 area()方法是 public 的，其属性 Pi 也是 final 和 static 的。

3. 在什么情况下,可以对超类对象的引用进行强制类型转换,使其转换成子类对象的引用?

解:

当某个超类对象的引用实质上指向一个子类对象时,可以转换为该子类对象的引用。

4. 声明一个接口,此接口至少具有一个方法;在一个方法中声明内部类实现此接口,并返回此接口的引用。

解:

新建 Exe4_4.java 文件,编写测试代码如下:

```
public interface Runnable {
    void run();
}
public class Main {
    public static void main(String[] args) {
        Runnable r = new Runnable(){
            @Override
            public void run() {
                System.out.println("run");
            }
        }
    }
}
```

5. 声明一个具有内部类的类,此内部类只有一个非默认的构造方法;声明另外一个具有内部类的类,此内部类继承第一个内部类。

解:

新建 Exe4_5.java 文件,编写测试代码如下。

```
class Outer1 {
    class Inner1 {
        private String name;
        Inner1 (String name) {
            this.name = name;
        }
    }
}

class Outer2 {
    class Inner2 extends Outer1.Inner1{
        Inner2 (String name) {
            super(name);
        }
    }
}
```

6. 声明一个具有两个方法的类，在第一个方法中调用第二个方法；声明此类的一个子类，并在子类中覆盖第二个方法；生成一个子类的对象，并将其类型转换为超类，调用第一个方法，解释会发生什么？

解：

新建 Exe4_6.java 文件，编写测试代码如下。

```
class A3 {
    void f1() {
        f2();
    }
    void f2() {
        System.out.println("In f2() of A3");
    }
}
class B3 extends A3{
    @Override
    void f2() {
        System.out.println("In f2() of B3");
    }
}
public class Exe4_4 {
    public static void main (String[] args) {
        B3 b = new B3();
        A3 a = (A3)b;
        a.f1();
    }
}
```

程序的输出结果如下。

```
In f2() of B3
```

由于对象 a 实质上是类 B3 的实例，因此调用的方法 f2()是经过覆盖的 f2()。

7. 什么是多态性？如何实现多态？

解：

将一条消息发给对象的引用时，可能并不知道运行时对象的具体类型会是什么，但采取的行动同样是正确的，这种情况就叫作多态性。多态性通过自动塑型实现。

8. 在第 3 章习题 11 的基础上，声明测试类完成对多态性的测试：

(1) 在主方法中声明 Student 类的数组(含 5 个元素)；

(2) 生成 5 个对象存入数组：其中 3 个 Student 类的对象，1 个 StudentXW 类的对象，1 个 StudentBZ 类的对象；

(3) 将方法 testScore()发送给数组的每一个元素，输出结果，并分析具体执行的是哪一个类中的方法。

解：

使用第3章习题11中的Student类、StudentXW类、StudentBZ类，并新建Exe4_8.java文件，其内容如下。

```
public class Exe4_8 {
    public static void main(String[] args) {
        Student[] students = new Student[5];
        students[0] = new Student(20100023, "张三", 85, 92, 90);
        students[1] = new Student(20100008, "肖三", 83, 80, 78);
        students[2] = new Student(20100012, "刘三", 92, 90, 93);
        StudentXW xw = new StudentXW(20100015, "李四", 80, 90, 95);
        students[3] = xw;
        StudentBZ bz = new StudentBZ(20100005, "王五", 82, 85, 88);
        students[4] = bz;
        //使用增强 for 循环遍历 students 数组
        for(Student s : students) {
            System.out.println(s.name + "的测评成绩为: " + s.testScore());
        }
    }
}
```

程序的输出结果如下。

```
张三的测评成绩为: 89
肖三的测评成绩为: 80
刘三的测评成绩为: 91
李四的测评成绩为: 91
王五的测评成绩为: 90
```

可以看到，对students[3]和students[4]，其testScore()方法分别是StudentXW类和StudentBZ类中的testScore()方法。

# 第 5 章

# 异常处理与输入输出流

## 要点导读

本章内容需要与配套的主教材《Java 语言程序设计》(第 3 版)第 5 章配合学习。

主教材第 5 章介绍了异常处理与输入输出流。异常(Exception)又称为例外,是特殊的运行错误对象。Java 提供了异常处理机制,在 Exception 类中定义了程序产生异常的条件。对待异常通常并不是简单地结束程序,而是转去执行某段特殊代码处理这个异常,设法恢复程序继续执行。Java 中预定义的常用异常有 ArithmeticException、NullPointerException、ArrayIndexOutOfBoundsException、NegativeArraySizeException、FileNotFoundException、IOException 等。对于异常,可以声明抛出,由调用方法处理,也可以捕获并处理。抛出异常使用 throws 子句实现,捕获和处理异常使用 try、catch 和 finally 语句。程序员也可以自定义异常类,表示某种异常。

在 Java 中输入/输出操作通常都是通过输入/输出流来实现的。一个流就是一个从源流向目的地的数据序列。输入/输出流可以与各种数据源和目标相连。Java 所有的输入和输出操作都要通过 IO 包中的一些流类的方法来实现。从流的方向划分,可以分为输入流和输出流;从流的分工划分,可以分为节点流和处理流;从流的内容划分,可以分为面向字符的流和面向字节的流。

java.io 包中提供了用于读写文本文件和二进制文件的类和方法,另外,对于大文件的读写,可以使用缓冲流提高效率。File 类是 IO 包中唯一表示磁盘文件信息的对象,它定义了一些与平台无关的方法来操纵文件。java.util.zip 包中提供了一些类,可以以压缩格式对流进行读写。java.io 包中提供了专门用于对象信息存储和读取的输入输出流类 ObjectInputStream 和 ObjectOutputStream。java.io 包提供了 RandomAccessFile 类用于随机文件的创建和访问。

## 实验 5　异常处理与输入输出流

### 一、实验目的

(1) 了解 Java 的异常处理机制,会编写异常处理程序。

(2) 理解输入输出(I/O)流的概念,掌握其分类。

(3) 掌握文本文件读写、二进制文件读写、处理流类的概念和用法、对象序列化。

（4）掌握 File 类、压缩流类、随机读写流类。

（5）遇到 I/O 方面的问题，能够自行查阅 API 文档解决。

## 二、实验任务

### 1. 函数返回及异常处理

目测以下程序代码能编译通过吗？不能的话有什么错误？若有错误则尝试在不改变原程序语义的情况下，给出两种解决方案并分别实现在 Calculator1.java 和 Calculator2.java 中，解决针对 Exception 的错误。找错误时先不使用 IDE 辅助，尽可能找出所有错误。

```
public class Calculator {

    public float getValue(String type) throws Exception {
        //获得产品单价
        if (type.equals("cookie")) {
            return 1.11f;
        } else if (type.equals("pie")) {
            return 5.5f;
        }
    }

    public int getValue(String type) throws Exception {
        //获得产品数量
        if (type.equals("cookie")) {
            return 10;
        } else {
            return 20;
        }
    }

    public float calculate() {
        float price = getValue("cookie");
        int amount = getValue("cookie");
        return price * amount;
    }
}
```

### 2. 商品找零

从键盘上输入一件商品的价格(0.01～5.00 元)。如果支付 5 元，给出一种找零方案，使得所找纸币及硬币的个数最少。例如，输入物品价格 1.68 元，给出找零方案 2 元 1 张，1 元 1 张，2 角 1 个，1 角 1 个，2 分 1 个。可找币种类包括所有发行的纸币和硬币。

如果输入在 0.01～5.00 范围内则给出方案并继续等待下个输入；如果输入 0，则程序退出。如果输入 3 位(含)以上的小数，例如 3.618，则抛出自定义异常 WrongFormatException，并让用户继续重新输入；如果输入负数或者大于 5 的数值，则抛出自定义异常

InputOutOfRangeException，并且程序终止。

**3. 各种输入/输出流的特点**

简要回答问题：FileReader 和 FileInputStream 有什么共同点？有什么区别？

读写文件时，可直接使用 FileReader/FileWriter，FileInputStream/FileOutputStream。也可以在其上套上一层 BufferedReader/BufferedWriter，BufferedInputStream/BufferedOutputStream。这两种方法有什么区别？一般读写较大文件时应该使用哪种方法？

写一行代码说明如何在 FileReader 上再套一层 BufferedReader。

**4. 继续完善记账软件：对象序列化**

在实验四的实验任务 3 基础上继续完善记账软件，增加对象序列化功能。

(1) 思考如何给 myaccount.model.Deal 实现对象序列化，序列化一个所有成员变量都不是 null 的 NormalDeal 对象(给它的所有成员变量赋值)，可以序列化成功吗？还需要对其他的类做什么修改？为什么？

(2) 保存二进制文件。

随机生成 10 个不同类型的具体 Deal 对象。要求类型、日期等变量为随机值。将它们以二进制的形式保存到文件中(deal.dat)。

(3) 保存文本文件。

将上题中保存的文件读入，并以一定格式显示到屏幕上，显示的同时将这些文本写入文本文件 dealreport.txt。

(4) 文件打包成 zip 文件。

编程将本任务中生成的 deal.dat 和 dealreport.txt 打包成 deal.zip 文件。

**5. 附加题：编码转换**

通过字节流类和字符流类，配合 byte[]和 String 类。实现一个文件到另一个文件的编码转换。例如，从 GBK 编码到 UTF-8 编码等的转换。了解 Java 相关的编码转换问题。

## 三、实验步骤

**1. 解决实验任务 1 中的编译错误**

(1) 目测题目中代码的编译错误。指出这些错误，并设计出两种异常的处理方法。

(2) 将两种解决方案实现。

**2. 验证实验任务 1 的程序**

(1) 创建项目 exp5，创建 Java 包，名称为“problem1”。

(2) 在 problem1 包下创建 Calculator.java。将题目中的代码输入，并编译。看看自己是否找到了所有的错误。编译器只显示最高层的错误，将高层的错误修改后可以看到更细节的错误。实际应该找到 3 个不同类型的编译错误。

(3) 在 problem1 包下创建 Calculator1.java 和 Calculator2.java。在不大量改变原程序语义的情况下，修改 Calculator.java 中的错误。分别将两种修改方案在这两个类中实现。

**3. 完成实验任务 2 商品找零**

(1) 在项目 exp5 中创建新的包和类。创建 Java 包，名称为“problem2”。在 problem2 包下创建 ChangeCalculator.java。

(2) 创建 WrongFormatException.java，InputOutOfRangeException.java。并实现这两个自定义异常类。

(3) 在 ChangeCalculator 类中实现题目中要求的方法。

(4) 对程序进行验证，输入以下数值验证程序的输出。

```
输入: 0          输出: 程序退出
输入: 5.00       输出: 无须找零的方案
输入: 1.68       输出: 2 元 1 张, 1 元 1 张, 2 角 1 个, 1 角 1 个, 2 分 1 个的方案
输入: 0.01       输出: 2 元 2 张, 5 角 1 个, 2 角 2 个, 5 分 1 个, 2 分 2 个的方案
输入: 2.222      输出: 显示异常 WrongFormatException 并等待重新输入
输入: 6.05       输出: 显示异常 InputOutOfRangeException 并退出
```

注意：此题最容易出现的错误是少找一分钱，例如，输入 0.01 时，输出的不是 2 个 2 分，而是 1 个 2 分和 1 个 1 分。其原因在于浮点数在计算机中的表示是不精确的。此题中，如果要知道某浮点数对应的整数，可以使该浮点数加上一个很小的数，并对结果取整。例如，对浮点数 8.99，如果想得到结果 9，可以使 8.99 加上 0.02，再取整，就可以得到整数 9，即

```
float f = 8.99;
int i = (int)(f + 0.02);
```

此时 i 的值为 9。

4. 完成改进版的记账软件

(1) 复习、理解流的概念、分类、功能，理解流的套接。思考如何给 myaccount.model.Deal 实现对象序列化。

(2) 打开 MyAccount 工程，将题目要求的代码实现在 myaccount.util 包下。

(3) 实现 RandomDealWriter 类。

(4) 实现 TextDealWriter 类。

(5) 实现 DealZipWriter 类。

5. 实现实验任务 5 的编码转换程序(附加，选做)

在目录 exp5 下创建 Java 包，名称为“problem5”。在 problem5 包下创建 EncodeTransfer.java，在其中完成实验任务 5。

(1) 使用 FileInputStream 类从源文件读取内容。

(2) 使用 String 类的 getBytes("GBK")方法将源文件内容解码。

(3) 使用 String 类的编码构造方法 String(xxx, "UTF-8")将内容进行重新编码。

(4) 使用 FileWriter 类将新编码输出到目标文件。

## 习题解答

1. 什么是异常？解释抛出异常和捕获异常的含义。

解：

异常(Exception)又称为例外，是特殊的运行错误对象，对应着 Java 语言特定的运行错

误处理机制。抛出是指：不在当前方法内处理异常，而是把异常抛出到调用方法中。捕获是指：使用 try{}catch(){}块，捕获到所发生的异常，并进行相应的处理。

2. 简述 Java 的异常处理机制。

解：

在一个方法的运行过程中，如果发生了异常，则这个方法(或者是 Java 虚拟机)便生成一个代表该异常的对象(包含该异常的详细信息)，并将它交给运行时系统，运行时系统查找方法的调用栈，从生成异常的方法开始进行回溯，直到找到包含相应异常处理的方法为止。

3. 系统定义的异常与用户自定义的异常有何不同？如何使用这两类异常？

解：

系统定义的异常是 Java 处理程序错误的一种方式，系统为可能产生非致命性错误的代码段设计错误处理模块，例如，整数除法中，除数为 0 等。用户自定义异常是为了防止程序中断或是出现未知错误。

4. 用户程序如何自定义异常？编程实现一个用户自定义异常。

解：

用户自定义异常类时，只需继承类 Exception 即可。

新建 Exe5_4.java 文件，其内容为：

```
class AgeBelowZeroExeption extends Exception{
    public AgeBelowZeroExeption() {
      super("Age is below zero");
   }
}
public class Exe5_4 {
    private int age;
    public static void setAge(int age) throws AgeBelowZeroExeption{
        if (age < 0)
            throw new AgeBelowZeroExeption();
    }
    public static void main(String[] args) {
        try {
            setAge(-2);
        }
        catch (Exception e) {
            System.out.println(e);
        }
    }
}
```

程序的输出结果如下。

```
AgeBelowZeroExeption: Age is below zero
```

5. 模仿文本文件复制的例题，编写对二进制文件进行复制的程序。

解：

新建 Exe5_5.java 文件，其内容为：

```
import java.io.*;
class BinaryFileCopy {  //声明一个类
    String sourceName, destName;
    FileInputStream source;
    FileOutputStream dest;
    String line;
    //这个私有方法用来打开源文件和目的文件，如无异常则返回 true
    private boolean openFiles() {
        try {
            source = new FileInputStream(sourceName);  //打开源文件
        }
        catch (IOException iox) {
            System.out.println("Problem opening " + sourceName);
                                                         //出现异常显示出错信息
            return false;
        }
        try {
            dest = new FileOutputStream(destName);      //打开目的文件
        }
        catch (IOException iox) {
            System.out.println("Problem opening " + destName);
            return false;
        }
        return true;
    }
    private boolean copyFiles() {  //这个私有方法用来复制文件，如无异常返回 true
        try {
            byte[] buf = new byte[512];
            int num = source.read(buf);          //从源文件读取数据
            while (num > 0) {                    //只要能够读取数据，就继续读
                dest.write(buf, 0, num);         //向目标文件写入
                num = source.read(buf);          //从源文件读取数据
            }
        }
        catch (IOException iox) {
            System.out.println("Problem reading or writing");
            return false;
        }
        return true;
    }
    private boolean closeFiles() {  //此私有方法用来关闭文件，如无异常返回 true
        boolean retVal = true;
```

```
        try {
            source.close();
        }
        catch (IOException iox) {
            System.out.println("Problem closing " + sourceName);
            retVal = false;
        }
        try {
            dest.close();
        }
        catch (IOException iox) {
            System.out.println("Problem closing " + destName);
            retVal = false;
        }
        return retVal;
    }
    public boolean copy(String src, String dst) {
    //这个类中唯一的公有方法,需两个字符串参数
        sourceName = src;
        destName = dst;
        //调用三个私有方法,若都正常返回 true,有问题则返回 false,并显示相应出错信息
        return openFiles() && copyFiles() && closeFiles();
    }
}
public class Exe5_5 {
    public static void main(String[] args) {    //main()函数为程序入口
        if (args.length == 2)         //要求提供两个参数作为源和目标文件名
        {
            new BinaryFileCopy().copy(args[0], args[1]);
        } //新建一个 CopyMaker 类的对象并执行其 copy()方法,参数由命令行提供
        else {
            //如果不是两个参数,则给出提示信息,程序结束
            System.out.println("Please Enter File names");
        }
    }
}
```

6. 创建存储若干随机整数的文本文件,文件名、整数的个数及范围均由键盘输入。

解:

使用 Keyboard 类获得输入,Keyboard 类如下。新建 Keyboard.java 文件,其内容为:

```
import java.util.Scanner;
import java.io.*;
public class Keyboard {
```

```
    static BufferedReader inputStream = new BufferedReader(new
InputStreamReader(System.in));
    public static int getInteger() {
        try {
            return (Integer.valueOf(inputStream.readLine().trim()).intValue());
        } catch (Exception e) {
            e.printStackTrace();
            return 0;
        }
    }
    public static String getString() {
        try {
            return inputStream.readLine();
        } catch (Exception e) {
            e.printStackTrace();
            return null;
        }
    }
}
```

新建 Exe5_6.java 文件，其内容为：

```
import java.io.*;
import java.util.*;
public class Exe5_6 {
    public static void main(String[] args) {
        System.out.println("请输入文件名：");
        String fileName = Keyboard.getString();
        System.out.println("请输入整数的个数：");
        int count = Keyboard.getInteger();
        System.out.println("请输入整数范围(最小)：");
        int rangeLow = Keyboard.getInteger();
        System.out.println("请输入整数范围(最大)：");
        int rangeHigh = Keyboard.getInteger();
        File f = new File(fileName);
        try {
            BufferedWriter bw = new BufferedWriter(new FileWriter(f));
            Random r = new Random();
            int range = rangeHigh - rangeLow;
            for(int i = 0; i < count; i++) {
                int number = r.nextInt(range);
                number += rangeLow;
                bw.write(((Integer)number).toString());
                bw.write("\r\n");
            }
```

```
            bw.close();
        }
        catch(Exception e) {
            System.out.println(e);
            System.exit(-1);
        }
    }
}
```

7. 分别使用 FileWriter 和 BufferedWriter 往文件中写入 10 万个随机数，比较用时的多少。

解：

新建 Exe5_7.java 文件，其内容为：

```
import java.io.*;
import java.util.*;
public class Exe5_7 {
    public static void main(String[] args) throws IOException {
        File f1 = new File("D:/random1.txt");
        FileWriter fw = new FileWriter(f1);
        Random r = new Random();
        Calendar start = Calendar.getInstance();
        for (int i = 0; i < 100000; i++) {
            int number = r.nextInt(100);
            fw.write(((Integer) number).toString());
        }
        fw.close();
        Calendar end = Calendar.getInstance();
        long time = end.getTimeInMillis() - start.getTimeInMillis();
        System.out.println("使用 FileWriter 的时间为" + time + "毫秒");
        File f2 = new File("D:/random2.txt");
        BufferedWriter bw = new BufferedWriter(new FileWriter(f2));
        start = Calendar.getInstance();
        for (int i = 0; i < 100000; i++) {
            int number = r.nextInt(100);
            bw.write(((Integer) number).toString());
        }
        bw.close();
        end = Calendar.getInstance();
        time = end.getTimeInMillis() - start.getTimeInMillis();
        System.out.println("使用 BufferedWriter 的时间为" + time + "毫秒");
    }
}
```

程序某次执行的结果为：

```
使用 FileWriter 的时间为 162 毫秒
使用 BufferedWriter 的时间为 15 毫秒
```

8. 用记事本程序创建一篇包含几十个英语单词的小文章，要求从屏幕输出每一个单词。

提示：查阅 StreamTokenizer、StringTokenizer 类的说明。

解：

新建 Exe5_8.java 文件，其内容为：

```
import java.io.*;
import java.util.*;
public class Exe5_8 {
    public static void main(String[] args) throws IOException {
        File f = new File("D:/word.txt");
        FileReader fr = new FileReader(f);
        char[] buf = new char[(int) f.length()];
        int num = fr.read(buf, 0, buf.length);
        String contents = new String(buf);
        StringTokenizer st = new StringTokenizer(contents);
        while (st.hasMoreTokens()) {
            String s = st.nextToken();
            System.out.println(s);
        }
    }
}
```

9. 从键盘输入一系列字母，将其存储到文件中，对其进行升序排序后，存到另一个文件中，并显示在屏幕上。

解：

新建 Exe5_9.java 文件，其内容为：

```
import java.io.*;
import java.util.*;
public class Exe5_9 {
    public static void main(String[] args) throws IOException{
        System.out.println("请输入字母序列:");
        String s = Keyboard.getString();
        File f1 = new File("D:/a.txt");
        FileWriter fw = new FileWriter(f1);
        fw.write(s);
        byte[] buf = s.getBytes();
        //进行升序排序
         for (int i = 0; i < buf.length - 1; i++) {
            for (int j = i + 1; j < buf.length; j++) {
```

```
                if (buf[i] > buf[j]) {
                    byte temp = buf[i];
                    buf[i] = buf[j];
                    buf[j] = temp;
                }
            }
        }
        fw.close();
        File f2 = new File("D:/b.txt");
        fw = new FileWriter(f2);
        s = new String(buf, 0, buf.length);//转化为 String
        fw.write(s);
        fw.close();
    }
}
```

10. 创建一学生类(包括姓名、年龄、班级、密码),创建若干该类的对象并保存在文件中(密码不保存),从文件读取对象后显示在屏幕上。

解:

新建 Exe5_10.java 文件,其内容为:

```
import java.io.*;
class Student implements Serializable {
    String name;
    int age;
    int theClass;
    transient String password;        //声明为 transient,表示不进行序列化
    Student(String name, int age, int theClass, String password) {
        this.name = name;
        this.age = age;
        this.theClass = theClass;
        this.password = password;
    }
    @Override
    public String toString() {
        return "姓名:" + name + "\t 年龄:" + age + "\t 班级:" + theClass
                + "\t 密码: " + password;
    }
}
public class Exe5_10 {
    public static void main(String[] args) throws IOException, ClassNotFoundException {
        Student[] students = new Student[3];
        students[0] = new Student("Zhao Bo", 20, 3, "passZhao");
        students[1] = new Student("Wang Li", 19, 2, "passWangTong");
```

```
        students[2] = new Student("Ma Tong", 21, 1, "passLi");
        ObjectOutputStream oos = new ObjectOutputStream(
              new FileOutputStream("students.dat"));  //创建一对象输出流
        for (Student s : students) {  //使用增强 for 循环向流中写对象
            oos.writeObject(s);
        }
        oos.close();                                  //关闭输出流
        Student[] students1 = new Student[3];
        ObjectInputStream ois = new ObjectInputStream(
              new FileInputStream("students.dat"));   //创建一对象输入流
        for (int i = 0; i < 3; i++) {
            //读入对象并强制转型为 Student 类
            students1[i] = (Student) ois.readObject();
            System.out.println(students1[i]);
        }
        ois.close();
    }
}
```

程序的输出结果如下。

```
姓名:Zhao Bo     年龄:20     班级:3     密码: null
姓名:Wang Li     年龄:19     班级:2     密码: null
姓名:Ma Tong     年龄:21     班级:1     密码: null
```

从输出结果可以看出,由于将 Student 类的 password 字段设为 transient,所以保存和读出 Student 对象时都不会对其进行处理。

11. 一家杂货店的店主,需要查询、输入、修改任何一件商品的品名、价格、库存量信息。请用随机存取文件满足其要求,可以更新、查询信息。每件商品的标志为其记录号。

解:

新建 Exe5_11.java 文件,其内容为:

```
import java.io.*;
class Goods {          //商品
    char name[] = {'\u0000', '\u0000', '\u0000', '\u0000',
        '\u0000', '\u0000', '\u0000', '\u0000'};
    //姓名字符数组,初始状态用 8 个 Unicode 编码的空格填满
    float price;       //价格
    int stock;         //库存量
    public Goods(String name, float price, int stock) throws Exception {
        if (name.toCharArray().length > 8) {   //如果字符长度大于 8,则只取前 8 个
            System.arraycopy(name.toCharArray(), 0, this.name, 0, 8);
        }
        else {                                 //字符长度小于 8,则有几个填几个
```

```
            System.arraycopy(name.toCharArray(), 0, this.name, 0,
                    name.toCharArray().length);
        }
        this.price = price;
        this.stock = stock;
    }
}
public class Exe5_11 {
    String Filename;
    public Exe6_10(String Filename) {  //构造函数,要求初始化随机读写的文件名
        this.Filename = Filename;
    }
    public void writeGoods(Goods goods, int n) throws Exception { //写第 n 条记录
        RandomAccessFile ra = new RandomAccessFile(Filename, "rw");
        ra.seek(n * 20);                    //将位置指示器移到指定位置上
        for (int i = 0; i < 8; i++) {
            ra.writeChar(goods.name[i]);
        }
        ra.writeFloat(goods.price);
        ra.writeInt(goods.stock);
        ra.close();
    }
    public void readGoods(int n) throws Exception {   //读第 n 条记录
        char buf[] = new char[8];
        RandomAccessFile ra = new RandomAccessFile(Filename, "r");
        ra.seek(n * 20);
        for (int i = 0; i < 8; i++) {
            buf[i] = ra.readChar();
        }
        System.out.print("name:");
        System.out.println(buf);
        System.out.println("price:" + ra.readFloat());
        System.out.println("stock:" + ra.readInt());
        ra.close();
    }
    public void updateGoods(int n, Goods goods) throws Exception{
        writeGoods(goods, n);
    }
    public static void main(String[] args) throws Exception {  //主函数
        Exe6_10 t = new Exe6_10("1.txt");
        Goods g1 = new Goods("剪刀", 10.0f, 20);                  //创建第一类商品
        Goods g2 = new Goods("肥皂", 3.5f, 100);                  //创建第二类商品
        Goods g3 = new Goods("钢笔", 22.5f, 50);                  //创建第三类商品
        t.writeGoods(g1, 0);        //写入第一类商品信息为第 0 条记录
        t.writeGoods(g3, 2);        //写入第三类商品信息为第 2 条记录
```

```
            System.out.println("查询第一类商品信息如下：");
            t.readGoods(0);             //读取第 0 条记录
            t.writeGoods(g2, 1);        //写入第二类商品信息为第 1 条记录
            System.out.println("查询第二类商品信息如下：");
            t.readGoods(1);             //读取第 1 条记录
            g2.price = 4.0f;
            t.updateGoods(1, g2);
            System.out.println("更新后,第二类商品信息如下：");
            t.readGoods(1);             //读取第 1 条记录
        }
    }
```

程序的输出结果如下。

```
查询第一类商品信息如下:
name:剪刀
price:10.0
stock:20
查询第二类商品信息如下:
name:肥皂
price:3.5
stock:100
更新后,第二类商品信息如下:
name:肥皂
price:4.0
stock:100
```

# 第 6 章

# 集合框架

## 要点导读

本章内容需要与配套的主教材《Java 语言程序设计》(第 3 版)第 6 章配合学习。

主教材第 6 章介绍了 Java 为组织同一类型的对象提供的集合框架。通过实现集合框架中的相关接口和类能够更好地管理群体对象,例如,对群体对象的遍历、排序、查找、比较等。流式 API 为数组、集合等批量数据提供了高效处理接口,例如,过滤、计数等。

数组的缺点是其大小自创建以后就固定了。如果需要在序列中存储不同类型的数据,或者需要动态改变其大小,就需要用集合类型。Java 中有很多与集合有关的接口及类,它们被组织在以 Collection 及 Map 接口为根的层次结构中,称为集合框架。集合框架中的接口有 Collection、Set、SortedSet、List、Map、SortedMap,常用的实现类有 HashSet、ArrayList、LinkedList、HashTable、HashMap 等。

Vector 和 ArrayList 都是实现了 Collection 接口的具体类。这两个类在应用中经常使用,都具有下面的功能:能够存储对象,不能存储基本类型的数据,除非将这些数据包裹在包裹类中;其容量能够根据空间需要自动扩充;增加元素方法的效率较高。

可以使用增强 for 循环来遍历集合类对象中的每一个元素。使用 Enumeration 或 Iterator 类会使遍历方法得到简化。

Map 接口为根的集合类用于存储"关键字"(key)和"值"(value)的元素对,其中每个关键字映射到一个值。哈希表也称为散列表,是用来存储群体对象的集合类结构,但是查找的速度很快。

## 实验 6　集合框架

### 一、实验目的

(1) 了解群体数据的组织以及 Java 的集合框架。

(2) 理解数组、Collection、Map 的区别,能够区分各自的使用范围以及优缺点。

(3) 掌握 Arrays 类、Vector 类、ArrayList 类、Enumeration 接口、Iterator 接口、HashTable 类、HashMap 类的常用方法。

## 二、实验任务

### 1. 复习 Java 中各种容器接口

图 6-1 是 Java 容器类的关系图，依图复习 Java 容器类的内容，尤其是黑框标出的常用的几个容器类的使用方法。简单总结一下 Collection、Iterator、List、Set、Map 这几个接口所代表的含义。

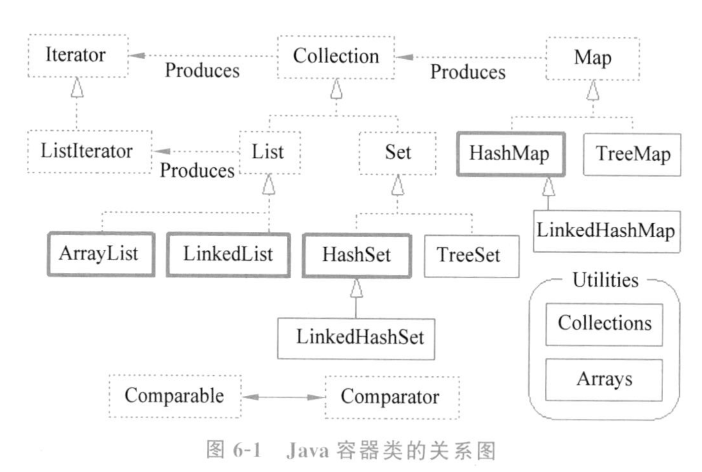


图 6-1　Java 容器类的关系图

### 2. HashMap 的使用

使用 HashMap 类重新实现实验三的字符串加密问题，不再使用 switch 或者 if-else 的方法实现，使用 HashMap 来保存密码字典。

原题如下。

实现 encode、decode 方法给字符串加密。输出原始字符串、加密后的字符串和解密后的字符串。

按照以下规律对测试字符串"We will break out of prison at dawn"进行替换加密，包括大小写字母。"_"代表空格字符。

a b c d e f g h i j k l m n o p q r s t u v w x y z _

v e k n o h z f _ i l j x d m y g b r c s w q u p t a

### 3. 继续完善记账软件：应用 ArrayList

在实验五的实验任务 4 基础上继续完善记账软件，生成并输出 10 个随机的 Deal 子类，amount 取值 0～1000。将它们放在 ArrayList 里，通过 Iterator 接口遍历这个 ArrayList，将 amount 小于 500 的 Deal 删除。再次遍历输出并检查结果。请考虑如何能正确删除。

### 4. 使用 TreeSet 类统计随机数

随机生成 80 个 1～100 的随机整数，输出一共生成了多少个不同的数，并按顺序输出曾出现过的数，相同的数只输出一次。

提示：使用 TreeSet 类。

## 三、实验步骤

### 1. 复习理解主教材第 6 章内容

复习、理解 Java 的容器类。

熟悉主教材第 6 章介绍的常用类的使用代码。

### 2. 重新实现实验 3 中的加密算法

（1）在工程 exp4 中新建包“problem2”，创建 StringEncryptor.java。

（2）在 StringEncryptor.java 中实现加密功能。（用 Hashtable 类的 put 操作保存字典，用 get 操作查询对应密文。）

### 3. 完成改进版的记账软件

将第 5 章的 MyAccount 包迁移到本章工程中，按实验任务 3 的要求，做如下改进。

（1）创建 myaccount.app.DealFilter 类。

（2）在 main()方法中测试题目功能。用 Deal 类指针的 ArrayList 保存不同 Deal 子类对象的地址，例如：

```
myDeal = (Deal)(new NormalDeal(…));
```

### 4. 实现用 TreeSet 类统计随机数

（1）创建项目“problem4”，添加类 RandomGenerator.java。

（2）在 RandomGenerator.java 用 TreeSet 类实现随机整数的去重保存。add()函数实现去重添加，for 循环默认为升序遍历。

# 习题解答

1. 数组的声明与数组元素的创建有什么关系？

解：

Java 在数组的声明中并不为数组元素分配内存，只有在数组元素创建时才为数组的元素分配内存。

2. Vector 类的对象与数组有什么关系？什么时候适合使用数组？什么时候适合使用 Vector？

解：

Vector 类的对象和数组都可以存储多个元素。当元素个数在创建时就确定时，适合使用数组。当元素个数开始不确定，在后面的处理过程中会变化时，适合使用 Vector。

3. 与顺序查找相比，二分查找有什么优势？使用二分查找有什么条件？

解：

顺序查找的算法简单，但在大数据量中进行查找时效率较低，对于已排序的数组，使用二分查找的效率比顺序查找要高。使用二分查找的条件是数组的元素已经排序。

4. 试举出三种常见的排序算法，并简单说明其排序思路。

解：

选择排序：基本思路是先在未排序序列中选一个最小元素，作为已排序子序列，然后再重复地从未排序子序列中选取一个最小元素，把它加到已经排序的序列中，作为已排序子序列的最后一个元素，直到把未排序子序列中的元素处理完为止。冒泡排序就是一种选择排序算法。

插入排序：基本思路是将待排序的数据按一定的规则逐一插入到已排序序列中的合适位置处，直到将全部数据都插入为止。其中最简单的为直接插入排序方法。

合并排序：基本思路是将数组分成两个子数组，分别排好序后，再合并成一个数组。对每个子数组的排序也使用同样的方法。

5. 声明一个类 People，成员变量有姓名、出生日期、性别、身高、体重等；生成 10 个 People 类对象，并放在一个一维数组中，编写方法按身高进行排序。

解：

新建 Exe6_5.java 文件，其内容为：

```
import java.util.Calendar;
//定义枚举类型
enum sex {
    MAIL,
    FEMAIL;
}
//定义 Birth 类，表示一个 Person 的出生日期
class Birth {
    int month;          //月份
    int day;            //年
    Birth(int month, int day) {
        this.month = month;
        this.day = day;
    }
}
class People {
    String name;
    Birth birth;
    sex s;
    float height;
    float weight;
    People(String name, Birth birth, sex s, float height, float weight) {
        this.name = name;
        this.birth = birth;
        this.s = s;
        this.height = height;
        this.weight = weight;
    }
    public String toString() {
        return this.name + "\t" + this.height;
    }
}
```

```
public class Exe6_5 {
    public static void main(String[] args) {
        //以下代码生成 10 个 Person 对象
        People[] p = new People[10];
        p[0] = new People("Wang Peng", new Birth(2, 15), sex.MAIL, 1.62f, 130f);
        p[1] = new People("Zhao Gang", new Birth(2, 5), sex.MAIL, 1.82f, 124f);
        p[2] = new People("Li Xin", new Birth(3, 1), sex.FEMAIL, 1.72f, 105f);
        p[3] = new People("Li Hua", new Birth(4, 22), sex.FEMAIL, 1.66f, 110f);
        p[4] = new People("Sun Jian", new Birth(8, 1), sex.MAIL, 1.75f, 120f);
        p[5] = new People("Wang Ze", new Birth(5, 15), sex.MAIL, 1.80f, 140f);
        p[6] = new People("Yang Bo", new Birth(3, 17), sex.MAIL, 1.76f, 138f);
        p[7] = new People("Li Hai", new Birth(10, 25), sex.MAIL, 1.65f, 132f);
        p[8] = new People("Liu Dong", new Birth(7, 29), sex.MAIL, 1.85f, 115f);
        p[9] = new People("Zhao Xun", new Birth(2, 30), sex.FEMAIL, 1.58f, 105f);
        //使用选择排序方法对 p 按照身高进行排序
        for (int i = 0; i < p.length; i++) {
            for (int j = i + 1; j < p.length; j++) {
                if (p[j].height > p[i].height) {
                    People temp = p[i];
                    p[i] = p[j];
                    p[j] = temp;
                }
            }
        }
        //使用增强 for 循环输出排序后的结果
        System.out.println("排序后的结果如下：");
        for (People people : p) {
            System.out.println(people);
        }
    }
}
```

程序的输出结果如下。

```
排序后的结果如下：
Liu Dong        1.85
Zhao Gang       1.82
Wang Ze         1.8
Yang Bo         1.76
Sun Jian        1.75
Li Xin          1.72
Li Hua          1.66
Li Hai          1.65
Wang Peng       1.62
Zhao Xun        1.58
```

6. 声明一个类，此类使用私有的 ArrayList 来存储对象。使用一个 Class 类的引用得到第一个对象的类型之后，只允许用户插入这种类型的对象。

解：

新建 Exe6_6.java 文件，其内容为：

```
import java.util.*;
class MyArrayList {
    ArrayList al;
    Class c;
    MyArrayList() {
        al = new ArrayList();
        c = null;
    }
    int add(Object o) {
        if(al.size() == 0) {      //al 为空时,记下第一个插入的对象的类型
            c = o.getClass();
            al.add(o);
            System.out.println("插入成功");
            return 0;
        }
        else if (o.getClass() != c) {
            System.out.println("类型"+o.getClass()+
                    "和前面输入的类型"+c+"不一样,插入失败");
            return -1;
        }
        else {
            al.add(o);
            System.out.println("插入成功");
            return 0;
        }
    }
}
public class Exe6_6 {
    public static void main(String[] args) {
        MyArrayList al = new MyArrayList();
        al.add(new Integer(3));
        al.add(new Float(2.5f));
        al.add(new Integer(5));
    }
}
```

程序的运行结果如下。

```
插入成功
类型 class java.lang.Float 和前面输入的类型 class java.lang.Integer 不一样,插入失败
插入成功
```

7. 找出一个二维数组的鞍点，即该位置上的元素在所在行上最大，在所在列上最小(也可能没有鞍点)。

解：

新建 Exe6_7.java 文件，其内容为：

```
import java.util.Random;
public class Exe6_7 {
    //查找数组 a 的鞍点的方法
    public static void findSaddle(int a[][]) {
        //maxi, maxj 分别记录鞍点的横坐标、纵坐标
        int maxi = 0; int maxj = 0; int flag = 0;
        for (int i = 0; i < a.length; i++) {
            int max = a[i][0];         //记录第 i 行中的最大值
            for (int j = 0; j < a[i].length; j++) {
                if (a[i][j] > max) {
                    maxi = i;
                    maxj = j;
                }
            }
            flag = 1;
            //对第 i 行中的最大值,判断是否是所在列的最小值
            for (int k = 0; k <a.length ; k++) {
                if (max > a[k][maxj]) {
                    flag = 0;
                    break;
                }
            }
            if (flag == 1) {
                break;
            }
        }
        System.out.println("maxi = " + maxi);
        System.out.println("maxj = " + maxj);
        if(flag == 1)
            System.out.println("鞍点是 a["+maxi+"]["+maxj+"]="+a[maxi][maxj]);
        else
            System.out.println("该二维数组没有鞍点");
    }
    public static void main(String[] args) {
        int M = 3, N = 3;
        int[][] a = new int[M][N];
        Random r = new Random();
        for(int i = 0; i < a.length; i++) {
            for(int j = 0; j < a[i].length; j++) {
```

```
                a[i][j] = r.nextInt(100);
            }
        }
        findSaddle(a);
    }
}
```

8. 声明一个矩阵类Matrix，其成员变量是一个二维数组，数组元素类型为int，设计下面的方法，并声明测试类对这些方法进行测试。

(1) 构造方法。

```
Matrix()                              //构造一个 10×10 个元素的矩阵,没有数据
Matrix(int n)                         //构造一个 n×n 个元素的矩阵,数据随机产生
Matrix(int n,int m)                   //构造一个 n×m 个元素的矩阵,数据随机产生
Matrix(int table[][])                 //以一个整型的二维数组构造一个矩阵
```

(2) 实例方法。

```
public void output()                  //输出 Matrix 类中数组的元素值
public Matrix transpose()             //求一个矩阵的转置矩阵
public Boolean isTriangular()         //判断一个矩阵是否为上三角矩阵
public Boolean isSymmetry()           //判断一个矩阵是否为对称矩阵
public void add(Matrix b)             //将矩阵 b 与接收者对象相加,结果放在接收者对象中
```

解：

新建Exe6_8.java文件，其内容为：

```
import java.util.*;
class Matrix {
    int n, m;
    int table[][];
    Matrix() {      //构造一个 10×10 个元素的矩阵,没有数据
        n = 10; m = 10;
        table = new int[n][m];
    }
    Matrix(int n) {      //构造一个 n×n 个元素的矩阵,数据随机产生
        this(n, n);
    }
    Matrix(int n, int m) {      //构造一个 n×m 个元素的矩阵,数据随机产生
        this.n = n;
        this.m = m;
        table = new int[n][m];
        Random r = new Random();
        for(int i = 0; i < n; i++) {
            for(int j = 0; j < m; j++) {
```

```
                table[i][j] = r.nextInt(100);
            }
        }
    }
    Matrix(int table[][]) {              //以一个整型的二维数组构造一个矩阵
        this.n = table.length;
        this.m = table[0].length;
        this.table = new int[n][m];
        for(int i = 0; i < n; i++)
            for(int j = 0; j < m; j++)
                this.table[i][j] = table[i][j];

    }
    public void output() {               //输出 Matrix 类中数组的元素值
        for (int[] a : table) {
            for (int b : a) {
                System.out.print(b + "\t");
            }
            System.out.println("");
        }
    }
    public Matrix transpose() {          //求一个矩阵的转置矩阵
        Matrix matrix = new Matrix(m, n);
        for (int i = 0; i < n; i++) {
            for (int j = 0; j < m; j++) {
                matrix.table[j][i] = table[i][j];
            }
        }
        return matrix;
    }
    public Boolean isTriangular() {  //判断一个矩阵是否为上三角矩阵
        boolean b = true;
        for (int i = 1; i < n; i++) {
            for (int j = 0; j < i; j++) {
                if (table[i][j] > 0) {
                    b = false;
                    break;
                }
            }
        }
        return b;
    }
    public Boolean isSymmetry() {  //判断一个矩阵是否为对称矩阵
        if(n != m)                    //如果行数和列数不等,则不是对称矩阵
            return false;
```

```
        boolean b = true;
        for (int i = 0; i < n; i++) {
            for (int j = i + 1; j < n; j++) {
                if (table[i][j] != table[j][i]) {
                    b = false;
                    break;
                }
            }
        }
        return b;
    }
    public void add(Matrix b) {      //将矩阵b与接收者对象相加,结果放在接收者对象中
        if (this.n != b.n || this.m != b.m) {
            System.out.println("两个矩阵的大小不一致,不能相加");
            return;
        } else {
            for (int i = 0; i < n; i++) {
                for (int j = 0; j < m; j++) {
                    table[i][j] += b.table[i][j];
                }
            }
        }
    }
}
//测试类
public class Exe6_8 {
    public static void main(String[] args) {
        int[][] table1 = { {1,2,3},{4,5,6},{7,8,9} };
        int[][] table2 = { {1,2,3},{0,4,5},{0,0,6} };
        Matrix m1 = new Matrix(table1);
        Matrix m2 = new Matrix(table2);
        if(m1.isTriangular())
            System.out.println("m1 是上三角矩阵");
        else
            System.out.println("m1 不是上三角矩阵");
        if(m1.isSymmetry())
            System.out.println("m1 是对称矩阵");
        else
            System.out.println("m1 不是对称矩阵");
        if(m2.isTriangular())
            System.out.println("m2 是上三角矩阵");
        else
            System.out.println("m2 不是上三角矩阵");
        if(m2.isSymmetry())
            System.out.println("m2 是对称矩阵");
```

```
        else
            System.out.println("m2 不是对称矩阵");
        Matrix m3 = m1.transpose();
        m1.add(m3); //m1 加上自己的转置矩阵,所以肯定是对称矩阵
        if(m1.isSymmetry())
            System.out.println("add(m3)后,m1 是对称矩阵");
        else
            System.out.println("add(m3)后,m1 不是对称矩阵");
    }
}
```

程序的运行结果如下。

```
m1 不是上三角矩阵
m1 不是对称矩阵
m2 是上三角矩阵
m2 不是对称矩阵
add(m3)后,m1 是对称矩阵
```

9. 用 key-value 对来填充一个 HashMap,并按 hash code 排列输出。

解:

新建 Exe6_9.java 文件,其内容为:

```
import java.util.*;
import java.util.Map.*;
public class Exe6_9 {
    public static void main(String[] args) {
        HashMap<String, String> hm = new HashMap<String, String>();
        hm.put("Mon", "Monday");
        hm.put("Tue", "Tuesday");
        hm.put("Wed", "Wednesday");
        hm.put("Thu", "Thursday");
        hm.put("Fri", "Friday");
        hm.put("Sat", "Saturday");
        hm.put("Sun", "Sunday");
        //使用 hm 的 toString()方法输出 hm 的内容
        System.out.println(hm);

        //创建 ArrayList 对象 lst,其内容为 hm 的所有 Entry
        ArrayList<Entry<String, String>> lst =
                new ArrayList<Entry<String, String>>(hm.entrySet());
        //对 lst 包含的所有对象排序,每一个对象为一个 Entry<String, String>
        //排序时,定义匿名类,实现 compare()方法
        Collections.sort(lst, new Comparator<Map.Entry<String, String>>() {
            @Override
```

```
            public int compare(Entry<String, String> o1,
                    Entry<String, String> o2) {
                return o1.hashCode() - o2.hashCode();
            }
        });
        //依次输出排好序后的 lst 的内容
        for (Entry<String, String> entry : lst) {
            System.out.println(entry.hashCode() + ":" + entry.getKey() + ":"
                    + entry.getValue());
        }
    }
}
```

程序的输出结果如下。

```
{Thu=Thursday, Sun=Sunday, Wed=Wednesday, Sat=Saturday, Fri=Friday, Tue=
Tuesday, Mon=Monday}
1636779419:Thu:Thursday
-1807270948:Sun:Sunday
-897522688:Wed:Wednesday
687262665:Tue:Tuesday
2112611714:Fri:Friday
-2049476577:Sat:Saturday
-1984564260:Mon:Monday
```

从输出结果可以看到，HashMap 的 toString()方法就是按照其中每个 Entry 的 hash code 从小到大依次输出每个 Entry。

10. 编写一个方法，在方法中使用 Iterator 类遍历 Collection，并输出此集合类中每个对象的 hashCode()值。用对象填充不同类型的 Collection 类对象，并将你的方法应用于每一种 Collection 类对象。

解：

新建 Exe6_10.java 文件，其内容为：

```
import java.util.*;
import java.util.Map.Entry;

public class Exe6_10 {
    //遍历 Collection 并输出 hash code 的方法
    public static void interateCollection(Collection c) {
        Iterator it = c.iterator();
        while (it.hasNext()) {
            Object o = it.next();
            System.out.println(o.hashCode());
        }
    }
```

```
    public static void main(String[] args) {
        Vector v = new Vector();
        v.add(1);
        v.add(1.2f);
        HashSet hs = new HashSet();
        hs.add(3.5d);
        hs.add("Hello");
        interateCollection(v);          //遍历 Vector 类的对象 v
        interateCollection(hs);         //遍历 HashSet 类的对象 hs
    }
}
```

程序的输出结果如下。

```
1
1067030938
69609650
1074528256
```

# 第 7 章

## 图形用户界面

### 要点导读

本章内容需要与配套的主教材《Java 语言程序设计》(第 3 版)第 7 章配合学习。

主教材第 7 章介绍了图形用户界面。GUI 组件的左上角坐标默认为(0,0),从左上角到右下角,水平坐标 x 和垂直坐标 y 增加。坐标的单位是像素,它是显示器分辨率的最小单位。通过指定坐标,文本和图形就可以显示在屏幕上指定的位置。

Java 中有关颜色的类是 Color 类,有关字体控制的类是 Font 类,使用 Graphics 类可以绘制线条、矩形、多边形、椭圆、弧等多种图形。Java2D API 提供了高级的二维图形功能。

在 Java 里用来设计 GUI 的组件和容器有两种,一种是早期版本的 AWT 组件,另一种是 Swing 组件,Swing 组件的名称都是在原来 AWT 组件名称前加上 J。Swing 组件提供了丰富、可扩展的 GUI 组件库,包括按钮、窗口、表格等一系列的图形组件。Swing 组件与 AWT 组件的最大不同之处在于 Swing 组件完全是由 Java 语言编写的,因此 Swing 组件的外观和功能不依赖于任何由宿主平台的系统所提供的代码,程序员可以为它设置在不同操作系统下统一的外观风格,当然也可以随操作系统的不同而变。

Swing 组件可以归为三个层次:顶层容器、中间层容器、原子组件。其中,顶层容器有三个,分别是 JFrame、JDialog 和 JApplet。中间层容器存在的目的仅仅是为了容纳别的组件,分为两类:一般用途的和特殊用途的。一般用途的有 JPanel、JScrollPane、JSplitPane、JTabbedPane、JToolBar 五类;特殊用途的有 JInternalFrame、JRootPane 两类。原子组件通常是在图形用户界面中和用户进行交互的组件。它的基本功能就是和用户交互信息。根据功能的不同,它可被分为三类:显示不可编辑信息的,如 JLabel、JProgressBar、JToolTip;有控制功能、可以用来输入信息的,如 JButton、JCheckBox、JRadioButton、JComboBox、JList、JMenu、JSlider、JSpinner、JTexComponent 等;能提供格式化的信息并允许用户选择的,如 JColorChooser、JFileChooser、JTable、JTree。

使用布局管理器可以指定每个组件的位置。使用布局管理器的方法是通过容器对象,调用其 setLayout()方法,并以某种布局管理器对象为参数。

编写事件处理程序时,要关注事件源、事件监听器、事件对象。事件源表示事件来自于哪个组件或对象。事件对象代表某个要被处理的事件。事件对象中包含事件的相关信息和事件源。事件监听器负责监听事件并做出响应,一旦它监视到事件发生,就会自动调用相应的事件处理程序做出响应。

桌面 API 允许 Java 应用程序完成以下三件事情:通过一个 URL,启用主机平台上默认的浏览器打开该URL,这个功能由DeskTop的browse()方法完成;启用主机平台上默认的邮件客户端,这个功能由DeskTop的mail()方法完成;对特定的文件,启用主机平台上与之关联的应用程序,对该文件进行打开、编辑、打印操作,这些功能分别由DeskTop的open()、edit()、print()方法完成。

# 实验7　图形用户界面

## 一、实验目的

(1) 掌握图形用户界面程序的编程方法、思路,学会在Application中引入图形用户界面。

(2) 了解Java的图形环境,绘制简单图形。

(3) 掌握Swing的结构和特点,学会使用布局管理、事件处理,以及常用的Swing组件。

## 二、实验任务

### 1. 复习图形用户界面的作用

图形用户界面的作用是什么?简要回答图形用户界面程序都需要负责处理哪些任务,需要实现哪些功能。

### 2. 使用Graphics类画图

至少使用Graphics类5个画图方法在Application窗口中绘制图形。

### 3. 使用Swing画图

实现一个Java Swing应用程序,至少使用Graphics2D类的5个画图方法,在窗口背景上画图。

### 4. Swing组件及简单事件响应练习

实现一个JFrame,其中包含JLabel、JTextField、JTextArea、JList、JButton等组件。合理地排列这些组件,如图7-1所示是一个简单的JFrame例子,能够响应“确定”按钮单击事件,单击后在右侧的TextArea中显示字符串,字符串的内容要包括TextField和List的输入值。

## 三、实验步骤

### 1. 复习图形用户界面的基本内容

理解、复习主教材中图形用户界面的相关知识点,掌握主要类的特点,熟悉主教材中相关代码。

### 2. 复习Graphics

(1) 新建“exp7”项目和“problem2”包。

(2) 参考主教材例7-1,实现实验任务2的要求。

### 3. 复习Java Swing以及Graphics2D

在目录exp7下建立problem3项目,参考主教材例7-2,使用Swing组件的Graphics2D类实现实验任务3的要求。

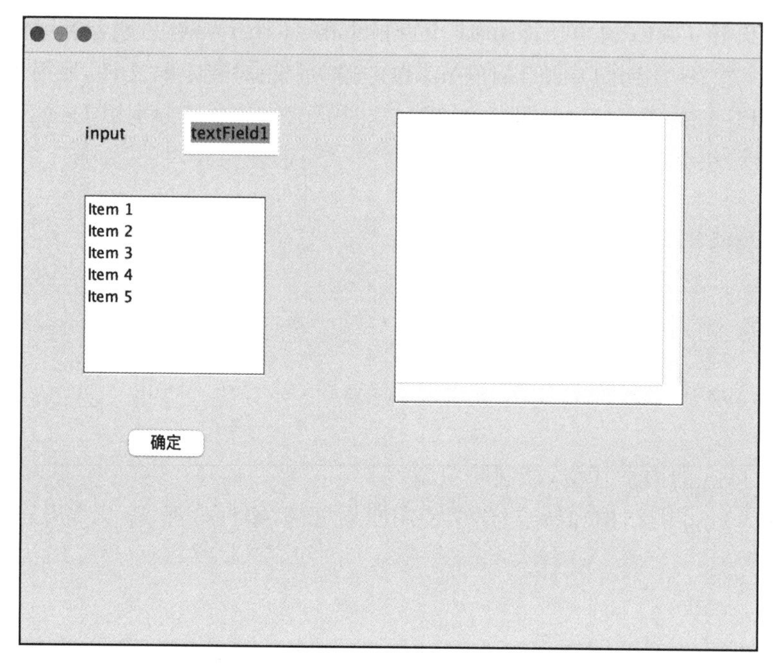


图 7-1　JFrame 的一个简单例子

4. 复习 Swing 组件及事件响应

(1) 在目录 exp7 中创建项目 problem4，按以下步骤实现实验任务 4 的要求。

(2) 右键单击 problem4 包，选择 New→Swing UI Designer→GUI Form 命令创建一个 javax.swing.JFrame 的子类以及它的 Form，如图 7-2 所示。

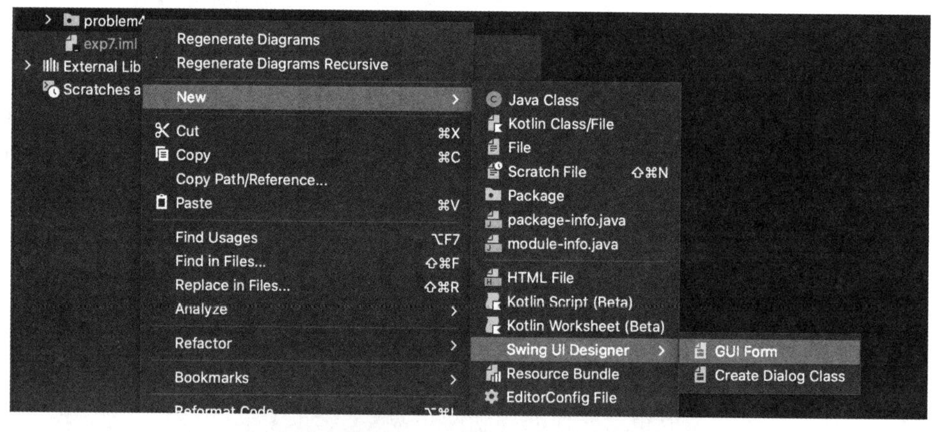


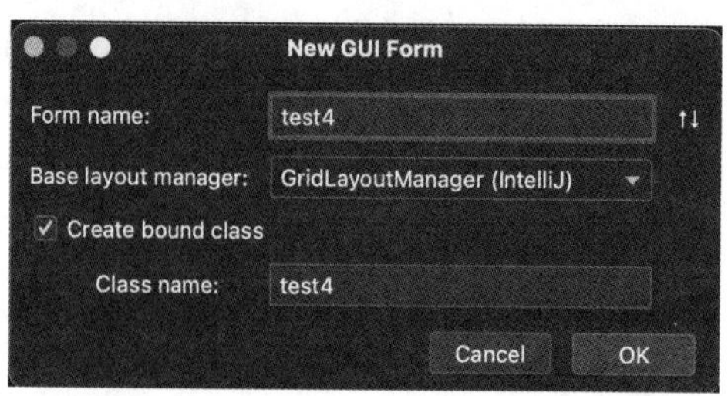


图 7-2　创建 Form

(3) 可以看到新建的 Java 类继承了 JFrame，它的图标也与一般的类不同。双击这个新建的 Java 文件，可以打开 JFrame 视图，如图 7-3 所示。

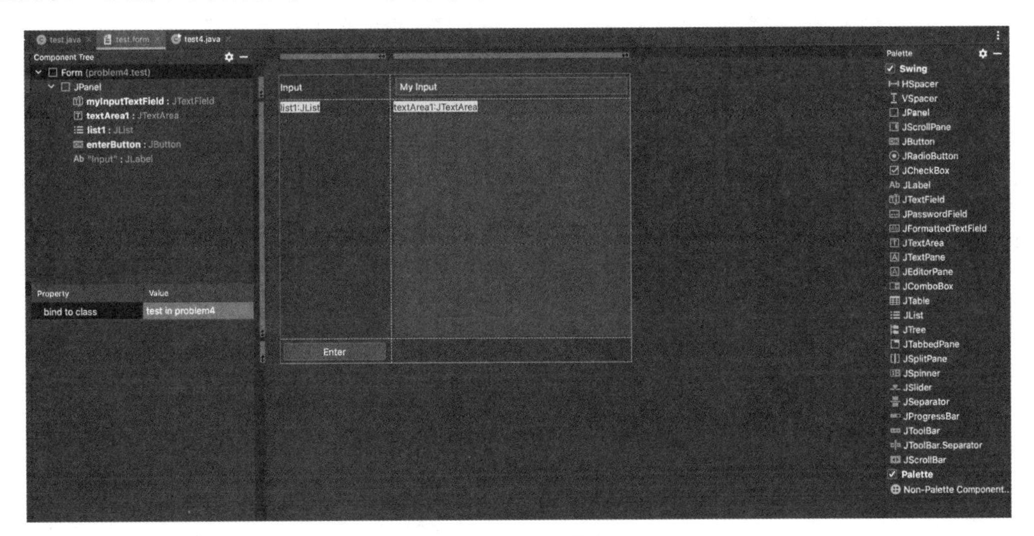

图 7-3　JFrame 的实现

(4) 左上角是这个 JFrame 所包含的所有组件结构；右面是所有 Swing 组件，可以通过拖曳来添加到 JFrame 中；左下是选中组件的属性和事件。

(5) 根据题目要求添加所需要的组件，并设置好它们的属性。为 JButton 组件添加 actionPerformed 事件，并在事件实现代码中填写题目要求的逻辑。

## 习题解答

1. 编写一个程序，该程序绘制一个 5×9 的网格，使用 drawLine 方法。

解：

新建 Exe7_1.java 文件，其内容为：

```
import java.awt.*;
import javax.swing.*;
public class Exe7_1 extends JFrame {
    public Exe7_1() {
        super("5x9 网格");                //调用父类构造方法，设置窗口标题
        setSize(400, 300);               //设置窗口初始大小
        setVisible(true);                //设置窗口为可见
    }
    @Override
    public void paint(Graphics g) {
        super.paint(g);
        int xStart = 100;                //第一条纵向线段的横坐标
        int xWidth = 20;                 //每个格子的宽度
```

```
        int xEnd = xStart + xWidth * 8;         //最后一条纵向线段的横坐标
        int yStart = 100;                       //第一条横向线的纵坐标
        int yWidth = 20;                        //每个格子的高度
        int yEnd = yStart + yWidth * 5;         //最后一条横向线的纵坐标
        //画纵向的线段,一共 10 条
        for(int x = xStart; x <= xEnd; x+=20) {
            g.drawLine(x, yStart, x, yEnd);
        }
        //画横向线段,一共 6 条
        for(int y = yStart; y <= yEnd; y+=20) {
            g.drawLine(xStart,y, xEnd, y);
        }
    }
    public static void main(String args[]) {
        Exe7_1 application = new Exe7_1();
        application.setDefaultCloseOperation(JFrame.EXIT_ON_CLOSE);
    }
}
```

程序的运行结果如图 7-4 所示。

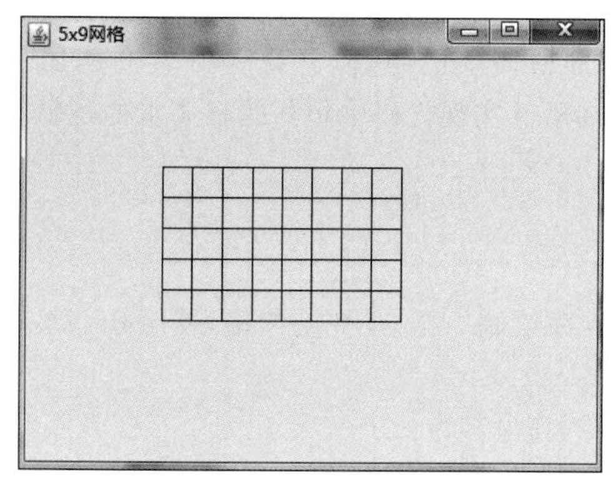


图 7-4　习题 1 程序运行结果

2. 编写一个程序,该程序以不同的颜色随机产生三角形,每个三角形用不同的颜色进行填充。

解:

新建 Exe7_2.java 文件,其内容为:

```
import java.awt.*;
import javax.swing.*;
import java.util.*;
public class Exe7_2 extends JFrame{
    public Exe7_2() {
        super("随机三角形");
```

```
        setSize(400, 300);
        setVisible(true);
    }
    @Override
    public void paint(Graphics g) {
        super.paint(g);
        int[] x = new int[3]; int[] y = new int[3];
        //画 5 个三角形,每个三角形顶点坐标随机产生
        for(int i = 0; i < 5; i++) {
            for(int j = 0; j < 3; j++) {
                //为了防止 5 个三角形的边有交叉,
                //设置每个三角形顶点的横坐标间隔较大
                x[j] = (int)(Math.random() * 50) + 100 + i * 50;
                y[j] = (int)(Math.random() * 50) + 100;
            }
            Random r = new Random();
            //随机生成颜色
            Color c = new Color(r.nextInt(256),
                    r.nextInt(256), r.nextInt(256));
            g.setColor(c);
            g.fillPolygon(x, y, 3);
        }
    }
    public static void main(String[] args) {
        Exe7_2 application = new Exe7_2();
        application.setDefaultCloseOperation(JFrame.EXIT_ON_CLOSE);
    }
}
```

程序的运行结果如图 7-5 所示。

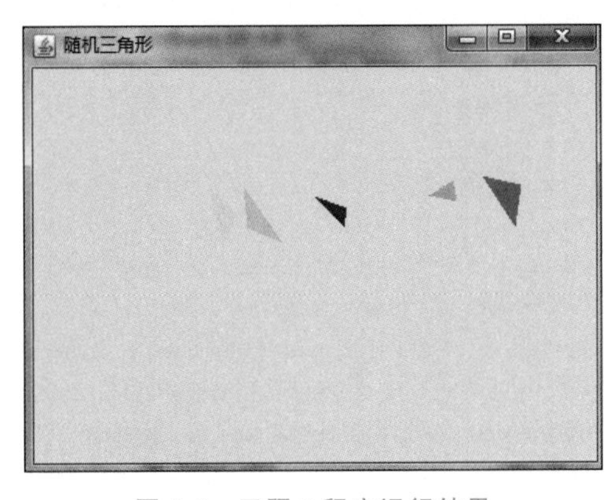


图 7-5　习题 2 程序运行结果

3. 编写一个绘制圆形的程序,当鼠标在绘制区域中单击时,该圆形的圆心应准确地跟随鼠标光标移动,重绘该圆形。

解：

新建 Exe7_3.java 文件，其内容为：

```
import java.awt.*;
import java.awt.event.*;
import javax.swing.*;
public class Exe7_3 extends JFrame{
    int centerX, centerY;
    double radius;
    public boolean isInCircle(int x, int y) {
        double dist = Math.sqrt((centerX - x) * (centerX - x) +
                (centerY - y) * (centerY - y));
        return dist <= radius;
    }
    public Exe7_3(int x, int y, double r) {
        super("随鼠标而动的圆");
        setSize(400, 300);
        setVisible(true);
        centerX = x;
        centerY = y;
        radius = r;
        this.addMouseListener(new MouseAdapter() {
            @Override
            public void mouseClicked(MouseEvent e) {
            int x = e.getX();int y = e.getY();

            if(isInCircle(x, y)) {
                centerX = x;
                centerY = y;
                repaint();
            }
        }});
    }
    @Override
    public void paint(Graphics g) {
        super.paint(g);
        Color c = Color.green;
        g.setColor(c);
        int leftUpX = centerX - (int)radius;
        int leftUpY = centerY - (int)radius;
        g.fillOval(leftUpX, leftUpY, (int)radius * 2, (int)radius * 2);
    }
    public static void main(String[] args) {
        Exe7_3 application = new Exe7_3(100, 100, 50);
        application.setDefaultCloseOperation(JFrame.EXIT_ON_CLOSE);
    }
}
```

程序运行时的初始界面如图 7-6 所示。

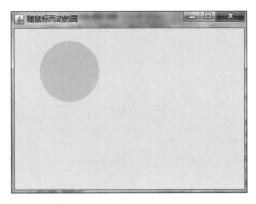

图 7-6　习题 3 程序运行初始界面

在圆所在的区域单击鼠标，圆心会移动到鼠标位置。

4. 编写一个"猜数"程序：该程序随机在 1～100 的范围内选择一个供用户猜测的整数，然后该程序显示提示信息，要求用户输入一个 1～100 的整数，根据输入数偏大、偏小、正确，程序将显示不同的图标。

解：

新建 Exe7_4.java 文件，其内容为：

```
import java.util.*;
import java.awt.*;
import java.awt.event.*;
import javax.swing.*;
public class Exe7_4 extends JFrame {
    int value = 0;
    public Exe7_4(int centerX, int centerY, double radius) {
        super("猜数游戏");
        setSize(400, 300);
        Container contentPane = getContentPane();
        GridLayout layout = new GridLayout(3, 1);
        layout.setVgap(100);
        contentPane.setLayout(layout);

        JButton button1 = new JButton("开始游戏");
        button1.addMouseListener(new MouseAdapter() {
            @Override
            public void mouseClicked(MouseEvent e) {
                Random r = new Random();
                value = r.nextInt(100);      //产生 0~99 的整数
                value = value + 1;           //变成 1~100 的整数
            }
        });
        contentPane.add(button1);

        JPanel centerPanel = new JPanel();
```

```
        centerPanel.setLayout(new GridLayout(1, 2));
        JLabel label1 = new JLabel("请输入您猜的数: ");
        final JTextField text = new JTextField();
        centerPanel.add(label1);
        centerPanel.add(text);
        contentPane.add(centerPanel);

        JButton button2 = new JButton("确定");
        button2.addMouseListener(new MouseAdapter() {
            @Override
            public void mouseClicked(MouseEvent e) {
                try {
                    int input = new Integer(text.getText()).intValue();
                    if (input == value) {
                        JOptionPane.showMessageDialog(null, "恭喜您,答对了");
                    } else if (input < value) {
                        JOptionPane.showMessageDialog(null, "太小");
                    } else {
                        JOptionPane.showMessageDialog(null, "太大");
                    }
                } catch (Exception exc) {
                    exc.printStackTrace();
                }
            }
        });
        contentPane.add(button2);
        setVisible(true);
    }
    public static void main(String[] args) {
        Exe7_4 application = new Exe7_4(50, 50, 50);
        application.setDefaultCloseOperation(JFrame.EXIT_ON_CLOSE);
    }
}
```

程序运行时的初始界面如图 7-7 所示。

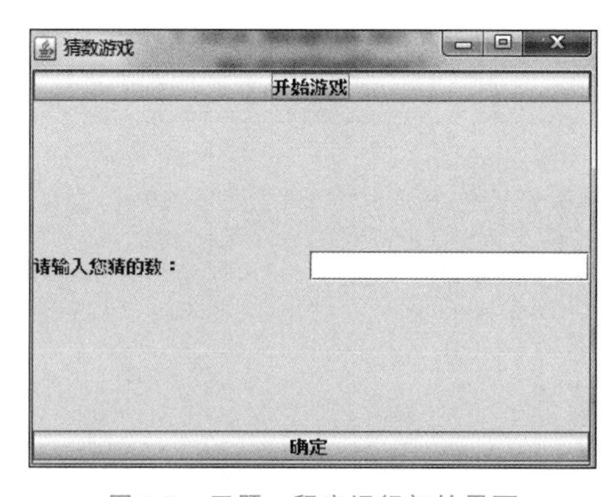


图 7-7　习题 4 程序运行初始界面

输入数字后会出现提示，猜对后会显示“猜对”。

5. 练习使用 JScrollPane。使用 BorderLayout 将 JFrame 布局分为左右两块；左边使用 GridLayout，包含 3 个按钮，右边在 JLabel 里显示一张图片，按钮控制 JLabel 是否显示滚动条。

解：

新建 Exe7_5.java 文件，其内容为：

```
import java.awt.*;
import java.awt.event.*;
import javax.swing.*;
public class Exe7_5 extends JFrame{
    JScrollPane scrollPane;
    Exe7_5 () {
        super("显示图片");
        setSize(400, 300);
        Container contentPane = getContentPane();
        contentPane.setLayout(new BorderLayout());

        JPanel leftPanel = new JPanel();
        leftPanel.setLayout(new GridLayout(3,1));
        JButton btn1 = new JButton("显示滚动条");
        JButton btn2 = new JButton("不显示滚动条");
        JButton btn3 = new JButton("适时显示滚动条");
        btn1.addMouseListener(new MouseAdapter() {
            @Override
            public void mouseClicked(MouseEvent e) {
                scrollPane.setHorizontalScrollBarPolicy(
                        JScrollPane.HORIZONTAL_SCROLLBAR_ALWAYS);
            scrollPane.setVerticalScrollBarPolicy(
                    JScrollPane.VERTICAL_SCROLLBAR_ALWAYS);
                scrollPane.revalidate();    //重新显示 JScrollPane 形状
            }
        });
        btn2.addMouseListener(new MouseAdapter() {
            @Override
            public void mouseClicked(MouseEvent e) {
                scrollPane.setHorizontalScrollBarPolicy(
                        JScrollPane.HORIZONTAL_SCROLLBAR_NEVER);
            scrollPane.setVerticalScrollBarPolicy(
                    JScrollPane.VERTICAL_SCROLLBAR_NEVER);
                scrollPane.revalidate();    //重新显示 JScrollPane 形状
            }
        });
        btn3.addMouseListener(new MouseAdapter() {
```

```
            @Override
            public void mouseClicked(MouseEvent e) {
                scrollPane.setHorizontalScrollBarPolicy(
                        JScrollPane.HORIZONTAL_SCROLLBAR_AS_NEEDED);
            scrollPane.setVerticalScrollBarPolicy(
                    JScrollPane.VERTICAL_SCROLLBAR_AS_NEEDED);
                scrollPane.revalidate();     //重新显示 JScrollPane 形状
            }
        });
        leftPanel.add(btn1);
        leftPanel.add(btn2);
        leftPanel.add(btn3);
        //左侧放 3 个按钮
        contentPane.add(leftPanel, BorderLayout.WEST);

        JLabel rightLabel = new JLabel(new ImageIcon("D:\\JAVA.jpg"));
        JPanel rightPanel = new JPanel();
        rightPanel.add(rightLabel);
        scrollPane = new JScrollPane(rightPanel);
        //中间放图片
        contentPane.add(scrollPane, BorderLayout.CENTER);
        setVisible(true);
    }
    public static void main(String[] args) {
        Exe7_5 application = new Exe7_5();
        application.setDefaultCloseOperation(JFrame.EXIT_ON_CLOSE);
    }
}
```

程序运行时的初始界面以及单击“不显示滚动条”后的界面如图 7-8 所示。

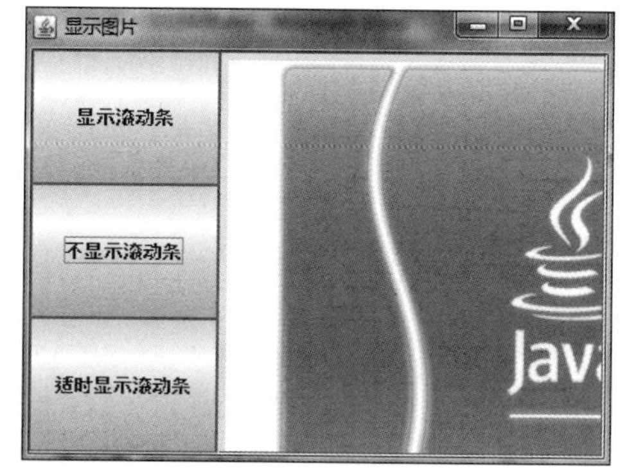


图 7-8　习题 5 程序运行界面

6. 练习使用 JList。建立两个 JList，双击其中任何一个中的某一项，此项就会添加到另外一个 JList 中。

解：

新建Exe7_6.java文件，其内容为：

```
import java.awt.*;
import java.awt.event.*;
import javax.swing.*;
public class Exe7_6 extends JFrame {
    JScrollPane scrollPane;
    DefaultListModel leftListModel;
    JList leftList;
    DefaultListModel rightListModel;
    JList rightList;
    Exe7_6() {
        super("JList示例");
        setSize(400, 300);
        Container contentPane = getContentPane();
        contentPane.setLayout(new GridLayout(1, 3));
        String[] data1 = {"one", "two", "three", "four"};
        leftListModel = new DefaultListModel();
        for (int i = 0; i < data1.length; i++) {
            leftListModel.addElement(data1[i]);
        }
        leftList = new JList(leftListModel);
        String[] data2 = {"five", "six", "seven"};
        rightListModel = new DefaultListModel();
        for (int i = 0; i < data2.length; i++) {
            rightListModel.addElement(data2[i]);
        }
        rightList = new JList(rightListModel);
        leftList.addMouseListener(new MouseAdapter() {
            @Override
            public void mouseClicked(MouseEvent e) {
                if (e.getClickCount() == 2) { //如果是双击事件
                    //将项目从leftList中remove,并加入rightList
                    int index = leftList.locationToIndex(e.getPoint());
                    String data = (String) leftListModel.getElementAt(index);
                    rightListModel.addElement(data);
                    rightList.setModel(rightListModel);
                    leftListModel.removeElementAt(index);
                    leftList.setModel(leftListModel);
                }
            }
        });
        rightList.addMouseListener(new MouseAdapter() {
            @Override
```

```
                public void mouseClicked(MouseEvent e) {
                    if (e.getClickCount() == 2) { //如果是双击事件
                        //将项目从 rightList 中 remove,并加入 leftList
                        int index = rightList.locationToIndex(e.getPoint());
                        String data = (String) rightListModel.getElementAt(index);
                        leftListModel.addElement(data);
                        leftList.setModel(leftListModel);
                        rightListModel.removeElementAt(index);
                        rightList.setModel(rightListModel);
                    }
                }
            });
            contentPane.add(new JScrollPane(leftList));
            contentPane.add(new JLabel());
            contentPane.add(new JScrollPane(rightList));
            setVisible(true);
        }
        public static void main(String[] args) {
            Exe7_6 application = new Exe7_6();
            application.setDefaultCloseOperation(JFrame.EXIT_ON_CLOSE);
        }
    }
```

程序运行时的初始界面如图 7-9 所示。

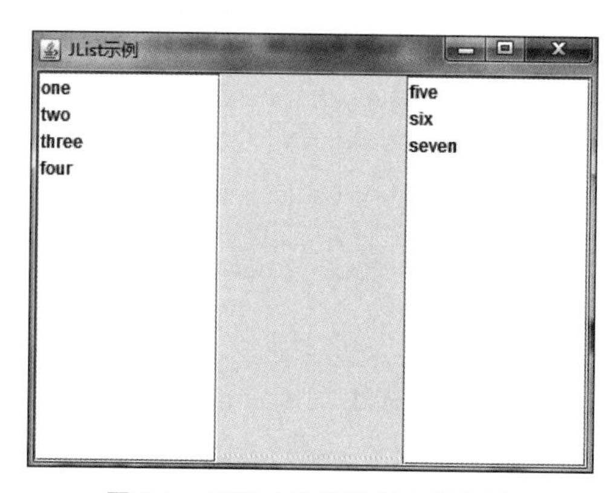


图 7-9　习题 6 程序运行初始界面

双击一栏中任一项会换到另一栏中。

7. 练习使用 JComboBox。包括一个 JLabel,一个 JComboBox,可以通过输入或者选择 JComboBox 中的某一项来控制 JLabel 中文字的大小。

解:

新建 Exe7_7.java 文件,其内容为:

```
import java.awt.*;
import java.awt.event.*;
```

```
import javax.swing.*;
public class Exe7_7 extends JFrame {
    String[] fontsize = {"4", "8", "12", "16", "20", "24", "28"};
    String defaultMessage = "请选择或直接输入字体大小!";
    Font font = null;
    JComboBox combo = null;
    JLabel label = null;
    ComboBoxEditor editor = null;
    public Exe7_7() {
        super("JComboBox示例");
        setSize(400, 300);
        Container contentPane = getContentPane();
        contentPane.setLayout(new BorderLayout());
        label = new JLabel("", JLabel.CENTER);
        font = new Font("SansSerif", Font.PLAIN, 12);
        label.setFont(font);
        combo = new JComboBox(fontsize);
        combo.setBorder(BorderFactory.createTitledBorder("请选择字体大小:"));
        combo.setEditable(true);
        editor = combo.getEditor();
        combo.configureEditor(editor, defaultMessage);
        combo.addItemListener(new ItemListener() {
            //实现 ItemListener 接口的 itemStateChanged()方法
            @Override
            public void itemStateChanged(ItemEvent e) {
                if (e.getStateChange() == ItemEvent.SELECTED) {
                    try {
                        int size = Integer.parseInt((String) e.getItem());
                        label.setText("目前选择的字形大小:" + size);
                    }
                    catch (Exception exc) {
                    }
                }
            }
        });
        combo.addActionListener(new ActionListener() {
            public void actionPerformed(ActionEvent e) {
                String s = (String) combo.getSelectedItem();
                try {
                    int size = Integer.parseInt(s);
                    font = new Font("SansSerif", Font.PLAIN, size);
                    label.setFont(font);
                }
                catch (Exception exc) {
                    editor.setItem("你输入的值不是整数值,请重新输入!");
```

```
                }
            }
        });
        contentPane.add(label, BorderLayout.CENTER);
        contentPane.add(combo, BorderLayout.SOUTH);
        setVisible(true);
    }
    public static void main(String[] args) {
        Exe7_7 application = new Exe7_7();
        application.setDefaultCloseOperation(JFrame.EXIT_ON_CLOSE);
    }
}
```

程序运行时的初始界面以及选择 20 号字体后的界面如图 7-10 所示。

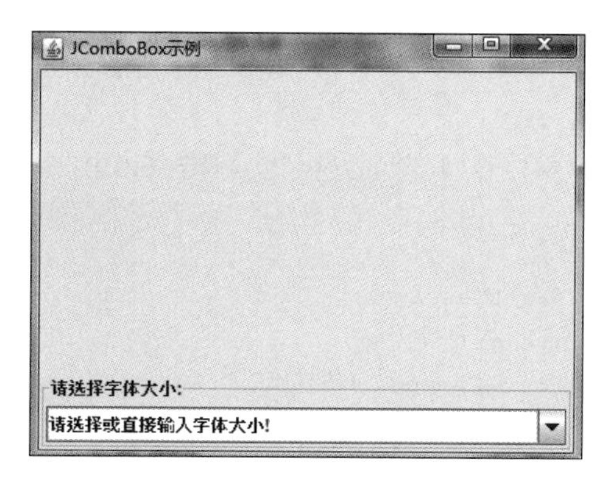

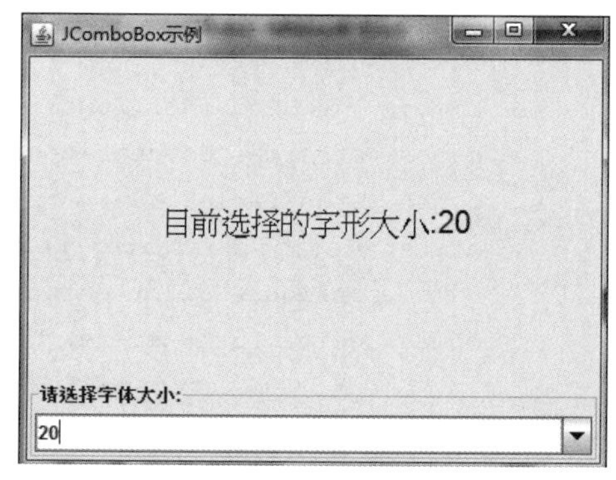


图 7-10　习题 7 程序运行界面

8. 练习使用 JTable。包括姓名、学号、语文成绩、数学成绩、总分五项，单击"总分"会自动将语文、数学成绩相加。

解：

新建 Exe7_8.java 文件，其内容为：

```
import javax.swing.*;
import javax.swing.table.*;
import java.awt.*;
import java.awt.event.*;
public class Exe7_8 extends JFrame {
    JTable table;
    DefaultTableModel model;
   Exe7_8() {
        super("JTable 示例");
        setDefaultCloseOperation(JFrame.EXIT_ON_CLOSE);
        //表格中显示的数据
        final Object rows[][] = {
            {"张三", "093235", "85", "90"},
            {"李四", "082898", "90", "88"},
```

```
        {"王五", "098928", "92", "86"}};
    final String columns[] = {"姓名", "学号", "语文成绩", "数学成绩", "总分"};
    model = new DefaultTableModel(rows, columns);
    table = new JTable(model);
    RowSorter sorter = new TableRowSorter(model);
    //设置排序
    table.setRowSorter(sorter);
    table.addMouseListener(new MouseAdapter() {
        @Override
        public void mouseClicked(MouseEvent e) {
            if (table.getSelectedColumn() != 4)
                return;
            //如果单击的是第 4 列
            int row = table.getSelectedRow();
            try {
              int sum = Integer.parseInt((String)model.getValueAt(row, 2))
                 + Integer.parseInt((String)model.getValueAt(row, 3));
              model.setValueAt(sum, row, 4);
              table.setModel(model);
            }
            catch (Exception exc) {
                exc.printStackTrace();
            }
        }
    });
    JScrollPane pane = new JScrollPane(table);
    add(pane, BorderLayout.CENTER);
    setSize(400, 300);
    setVisible(true);
}
public static void main(String args[]) {
    Exe7_8 application = new Exe7_8();
    application.setDefaultCloseOperation(JFrame.EXIT_ON_CLOSE);
}
}
```

程序运行时的初始界面以及计算总分后的界面如图 7-11 所示。

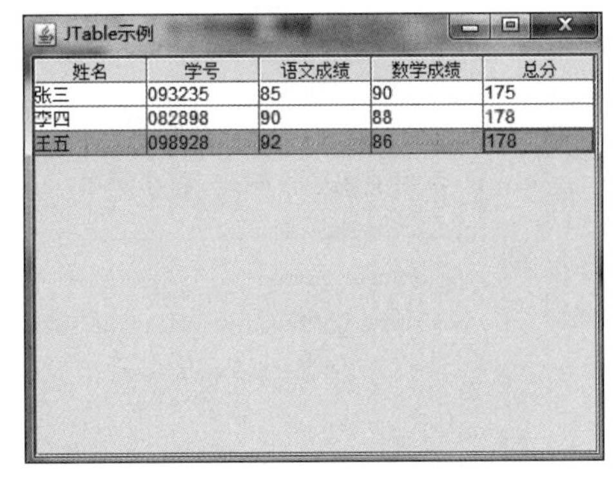


图 7-11　习题 8 程序运行界面

9. 练习使用对话框。包括一个 JLabel 和两个 JButton，单击任何一个 JButton 都会产生一个输入对话框，单击“确定”按钮后将输入内容在 JLabel 中显示出来。

解：

新建 Exe7_9.java 文件，其内容为：

```
import javax.swing.*;
import java.awt.*;
import java.awt.event.*;
public class Exe7_9 extends JDialog{
    JLabel label;
    JButton btn1, btn2;
    Exe7_9() {
        setSize(400, 300);
        Container contentPane = getContentPane();
        contentPane.setLayout(new GridLayout(2,1));

        label = new JLabel();
        contentPane.add(label);
        JPanel panel = new JPanel();
        btn1 = new JButton("按钮 1");
        btn2 = new JButton("按钮 2");
        //两个按钮的鼠标事件响应,弹出输入对话框获得输入
        MouseAdapter mAdapter = new MouseAdapter() {
            @Override
            public void mouseClicked(MouseEvent e) {
                String input = JOptionPane.showInputDialog("请输入");
                label.setText(input);
            }
        };
        btn1.addMouseListener(mAdapter);
        btn2.addMouseListener(mAdapter);
        panel.setLayout(new GridLayout(1,2));
        panel.add(btn1);
        panel.add(btn2);
        contentPane.add(panel);
        pack();
        setVisible(true);
        //添加关闭事件响应,程序退出
        this.addWindowListener(new WindowAdapter(){
            @Override
            public void windowClosing(WindowEvent e) {
                System.exit(0);
            }
        });
```

```
    }
    public static void main(String args[]) {
        Exe7_9 application = new Exe7_9();
        application.setDefaultCloseOperation(JDialog.DO_NOTHING_ON_CLOSE);
    }
}
```

程序运行时的初始界面、输入对话框界面、得到输入后的界面如图 7-12 所示。

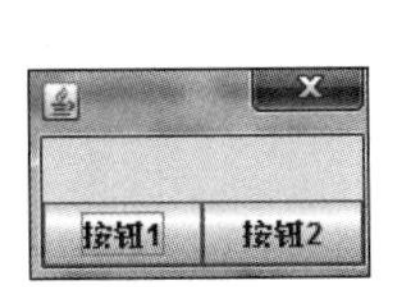

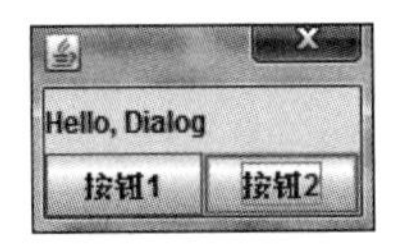


图 7-12　习题 9 程序运行界面

10. 练习使用 JMenu、JFileChooser、JColorChooser。通过菜单可以打开文件选择对话框，然后打开某一指定文本文件，通过菜单可打开颜色选择对话框控制显示文本的颜色。

解：

新建 Exe7_10.java 文件，其内容为：

```
import java.io.*;
import java.awt.*;
import java.awt.event.*;
import javax.swing.*;
public class Exe7_10 extends JFrame {
    Container frameContainer;
    JMenuBar menuBar = new JMenuBar();
    JMenu fileMenu = new JMenu("操作");
    JMenuItem fileItem = new JMenuItem("打开文件");
    JMenuItem colorItem = new JMenuItem("选择颜色");
    JMenuItem exitItem = new JMenuItem("退出");
    public Exe7_10() {
        super("菜单示例");
        fileMenu.add(fileItem);
        fileMenu.add(colorItem);
        fileMenu.add(exitItem);
        menuBar.add(fileMenu);
        setJMenuBar(menuBar);
        frameContainer = getContentPane();
        frameContainer.setLayout(null);
        fileItem.addActionListener(new MenuItemHandler());
        colorItem.addActionListener(new MenuItemHandler());
        exitItem.addActionListener(new MenuItemHandler());
        setSize(400, 300);
        setVisible(true);
    }
```

```
    public class MenuItemHandler implements ActionListener {
        public void actionPerformed(ActionEvent e) {
            String cmd = e.getActionCommand();
            if (cmd.equals("退出")) {
                System.exit(0);
            }
            else if (cmd.equals("打开文件")) {
                JFileChooser fChooser = new JFileChooser("E:\\..\\..");
                fChooser.setDialogTitle("打开文件");
                int result = fChooser.showOpenDialog(null);
                if (result == JFileChooser.APPROVE_OPTION) { //确认打开
                    File fileIn = fChooser.getSelectedFile();
                    JOptionPane.showMessageDialog(null,
                        "您选择的文件为: " + fileIn.getAbsolutePath());
                }
            }
            else if (cmd.equals("选择颜色")) {
                Color selectedColor = JColorChooser.showDialog(null,
                        "JColorChooser Sample", Color.RED);
                if (selectedColor != null) {
                    JOptionPane.showMessageDialog(null,
                            "您选择的颜色为:[" + selectedColor.getRed()
                            + "," + selectedColor.getGreen() + ","
                            + selectedColor.getBlue() + "].");
                }
            }
        }
    }
    public static void main(String[] args) {
        Exe7_10 application = new Exe7_10();
        application.setDefaultCloseOperation(JFrame.EXIT_ON_CLOSE);
    }
}
```

程序运行时的初始界面如图 7-13 所示。

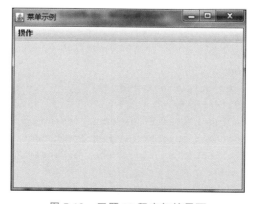

图 7-13　习题 10 程序初始界面

选择菜单“操作”→“打开文件”，弹出“打开文件”对话框，如图 7-14 所示。

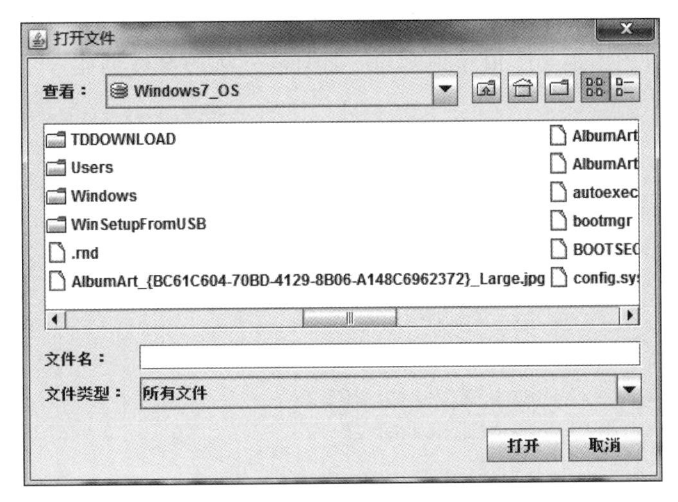


图 7-14　“打开文件”对话框

选择某文件，单击“打开”按钮后，将弹出文件绝对路径的消息框，如图 7-15 所示。

图 7-15　文件绝对路径消息框

选择某颜色，单击“确定”按钮后，将弹出选择颜色 RGB 值的消息框，如图 7-16 和图 7-17 所示。

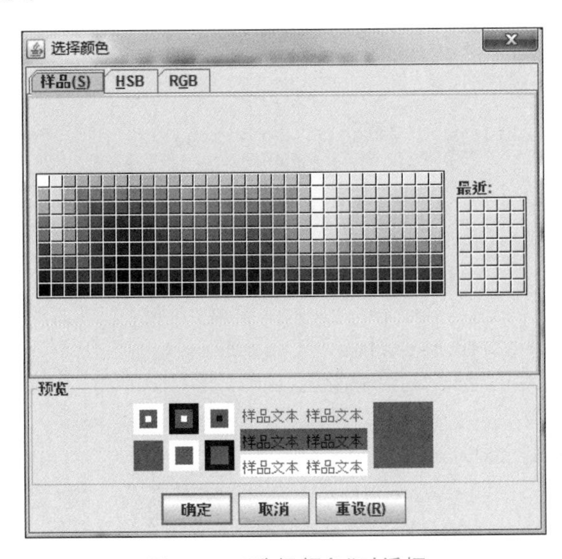


图 7-16　“选择颜色”对话框

图 7-17　选择颜色后的消息框

11. 编写一个图形用户界面，包括 3 个 JSlider 对象和 3 个 JTextField 对象。每个 JSlider 代表颜色中的红、绿、蓝 3 部分，它们的值为 0～255，在相应的 JTextField 中显示各个 JSlider 的当前值。用这 3 个值作为 Color 类构造方法的参数创建一个新的 Color 对象，用来填充一个矩形。

解：

新建 Exe7_11.java 文件，其内容为：

```
import java.awt.*;
import javax.swing.*;
import javax.swing.event.*;
public class Exe7_11 extends JFrame {
    JSlider rSlider = new JSlider(0, 255);
    JSlider gSlider = new JSlider(0, 255);
    JSlider bSlider = new JSlider(0, 255);
    JTextField rText = new JTextField();
    JTextField gText = new JTextField();
    JTextField bText = new JTextField();
    JPanel panel1;
    JPanel panel2;
    Exe7_11() {
        super("JSlider示例");
        Container contentPane = getContentPane();
        contentPane.setLayout(new GridLayout(2, 1));
        panel1 = new JPanel();
        panel2 = new JPanel();
        contentPane.add(panel1);
        contentPane.add(panel2);
        panel1.setLayout(new GridLayout(3, 2));
        panel1.add(rSlider);
        panel1.add(rText);
        rText.setText(new Integer(rSlider.getValue()).toString());
        panel1.add(gSlider);
        panel1.add(gText);
        gText.setText(new Integer(gSlider.getValue()).toString());
        panel1.add(bSlider);
        panel1.add(bText);
        bText.setText(new Integer(bSlider.getValue()).toString());
        class SliderListener implements ChangeListener {
            @Override
            public void stateChanged(ChangeEvent e) {
                JSlider js = (JSlider) e.getSource();  //得到js对象
                if (!js.getValueIsAdjusting())          //判断是否还在移动滑条
                {
                    if (js == rSlider) {
```

```
                    rText.setText(new Integer(js.getValue()).toString());
                } else if (js == gSlider) {
                    gText.setText(new Integer(js.getValue()).toString());
                } else if (js == bSlider) {
                    bText.setText(new Integer(js.getValue()).toString());
                }
                fillRect();
            }
        }
    }
    rSlider.addChangeListener(new SliderListener());
    gSlider.addChangeListener(new SliderListener());
    bSlider.addChangeListener(new SliderListener());
    setSize(400, 300);
    setVisible(true);
  }
  public void fillRect() {
    Graphics g = panel2.getGraphics();
    g.setColor(new Color(rSlider.getValue(),
            gSlider.getValue(), bSlider.getValue()));
    g.fillRect(150, 50, 100, 40);
  }
  public static void main(String[] args) {
    Exe7_11 application = new Exe7_11();
    application.setDefaultCloseOperation(JFrame.EXIT_ON_CLOSE);
  }
}
```

程序运行时的界面如图 7-18 所示。

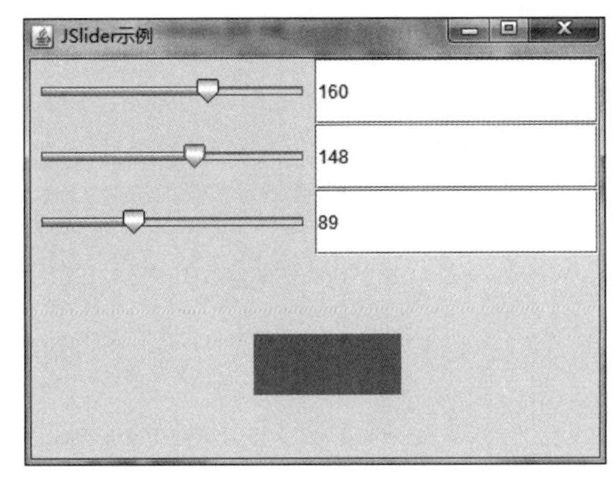


图 7-18　习题 11 程序运行界面

12. 编写 Application 程序，构造一 GUI，实现对两个数的加、减、乘、除功能。应含有 3 个 JTextField，一个 JButton。3 个 JTextField 分别用于输入两个数字和运算符号，结果用 JLabel 显示出来。

解：

新建 Exe7_12.java 文件，其内容为：

```
import java.awt.*;
import java.awt.event.*;
import javax.swing.*;
public class Exe7_12 extends JFrame {
    JTextField text1, text2, text3;
    JPanel panel;
    JLabel label;
    JButton btn;
    Exe7_12() {
        Container contentPane = getContentPane();
        GridLayout totalLayout = new GridLayout(3, 1);
        totalLayout.setHgap(50);
        totalLayout.setVgap(50);
        contentPane.setLayout(totalLayout);
        text1 = new JTextField();
        text2 = new JTextField();
        text3 = new JTextField();
        panel = new JPanel();
        GridLayout lay = new GridLayout(1, 3);
        lay.setHgap(2);
        panel.setLayout(lay);
        panel.add(text1);
        panel.add(text2);
        panel.add(text3);
        label = new JLabel();
        btn = new JButton("计算");
        contentPane.add(panel);
        contentPane.add(label);
        contentPane.add(btn);
        //按钮的处理函数,计算结果
        btn.addMouseListener(new MouseAdapter() {
            @Override
            public void mouseClicked(MouseEvent e) {
                char operator;
                double operand1;
                double operand2;

                try {//捕获异常,处理输入的操作数不符合要求的情况
                    operand1 = Double.parseDouble(text1.getText());
                    operand2 = Double.parseDouble(text3.getText());
                }
                catch(Exception exc) {
```

```
                JOptionPane.showMessageDialog(null, "请输入正确的操作数");
                return;
            }
            //获得操作符
            if (text2.getText().length() != 1) {
                JOptionPane.showMessageDialog(null, "请输入正确的操作符");
                return;
            }
            operator = text2.getText().charAt(0);
            //计算结果
            double result = 0.0;
            if (operator == '+') {
                result = operand1 + operand2;
            } else if (operator == '-') {
                result = operand1 - operand2;
            } else if (operator == '*') {
                result = operand1 * operand2;
            } else if (operator == '/') {
                if (operand2 == 0) {
                    JOptionPane.showMessageDialog(null, "除数不能为 0");
                    return;
                } else {
                    result = operand1 / operand2;
                }
            } else {
                JOptionPane.showMessageDialog(null, "请输入正确的操作符");
            }
            //显示结果
            label.setText("结果为:" + new Double(result).toString());
        }
    });
    setSize(400, 300);
    setVisible(true);
  }
  public static void main(String[] args) {
    Exe7_12 application = new Exe7_12();
    application.setDefaultCloseOperation(JFrame.EXIT_ON_CLOSE);
  }
}
```

程序运行时的界面如图 7-19 所示。

13. 编写一个含菜单的应用程序，包含 File 和 Type 两个菜单，File 菜单中包括“打开”和“退出”两个选项，打开菜单会弹出一个 JFileChooser 对话框，Type 菜单包含一系列复选框，可用于确定文件选择对话框的选择类型。

图 7-19　习题 12 程序运行界面

解：

新建 Exe7_13.java 文件，其内容为：

```
import java.awt.event. * ;
import java.io. * ;
import javax.swing.filechooser. * ;
import javax.swing. * ;
import java.util. * ;
public class Exe7_13 extends JFrame {
    JMenuBar menuBar = new JMenuBar();
    JMenu fileMenu = new JMenu("File");
    JMenuItem fileOpen = new JMenuItem("打开");
    JMenuItem fileExit = new JMenuItem("退出");
    JMenu editMenu = new JMenu("Type");
    JCheckBoxMenuItem[] typeItem = new JCheckBoxMenuItem[2];
    String[] name = new String[2];
    ButtonGroup buttonGroup = new ButtonGroup();
    public Exe7_13() {
        super("混合菜单示例");
        fileMenu.add(fileOpen);
        fileMenu.add(fileExit);
        typeItem[0] = new JCheckBoxMenuItem("* .jpg");
        typeItem[1] = new JCheckBoxMenuItem("* .txt");
        //每个 typeItem 代表的文件类型扩展名
        name[0] = new String("jpg");
        name[1] = new String("txt");
        for (int i = 0; i < typeItem.length; i++) {
            editMenu.add(typeItem[i]);
        }
        menuBar.add(fileMenu);
        menuBar.add(editMenu);
```

```
        setJMenuBar(menuBar);          //设置菜单
        //设置事件处理
        fileOpen.addActionListener(new MenuItemHandler());
        fileExit.addActionListener(new MenuItemHandler());
        setSize(400, 300);
        setVisible(true);
    }
    public static void main(String[] args) {
        Exe7_13 application = new Exe7_13();
        application.setDefaultCloseOperation(JFrame.EXIT_ON_CLOSE);
    }
    public class MenuItemHandler implements ActionListener {
        public void actionPerformed(ActionEvent e) {
            String cmd = e.getActionCommand();
            if (cmd.equals("退出")) {
                System.exit(0);
            }
            else if (cmd.equals("打开")) {
                JFileChooser fChooser = new JFileChooser("E:\\..\\..");
                String description = new String();
                Vector<String> selectedNames = new Vector<String>();
                //先统计已经选上的文件扩展名
                for (int i = 0; i < typeItem.length; i++) {
                    if (typeItem[i].isSelected()) {
                        selectedNames.add(name[i]);
                    }
                }
                if (selectedNames.size() > 0) {
                    String[] ext = new String[selectedNames.size()];
                    for (int i = 0; i < selectedNames.size() - 1; i++) {
                        //文件打开对话框的描述符
                        description = description + name[i] + "&";
                        //文件扩展名数组
                        ext[i] = selectedNames.elementAt(i);
                    }
                    description = description + selectedNames.lastElement();
                    ext[selectedNames.size() - 1] = selectedNames.lastElement();
                    //构造一个 FileNameExtensionFilter 对象,注意使用了可变长参数
                    FileNameExtensionFilter filter =
                            new FileNameExtensionFilter(description, ext);
                    fChooser.setFileFilter(filter);
                }
                fChooser.setDialogTitle("打开文件");
                int result = fChooser.showOpenDialog(null);
                if (result == JFileChooser.APPROVE_OPTION) {    //确认打开
```

```
                File fileIn = fChooser.getSelectedFile();
                JOptionPane.showMessageDialog(null,
                        "您选择的文件为:" + fileIn.getAbsolutePath());
            }
        }
    }
  }
}
```

程序运行时的初始界面如图 7-20 所示。

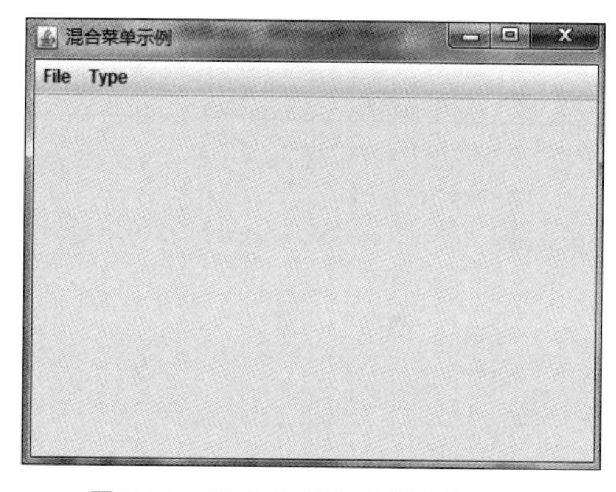


图 7-20　习题 13 程序运行初始界面

在 type 菜单中勾选"＊.jpg"和"＊.txt"后，出现的文件选择对话框如图 7-21 所示。

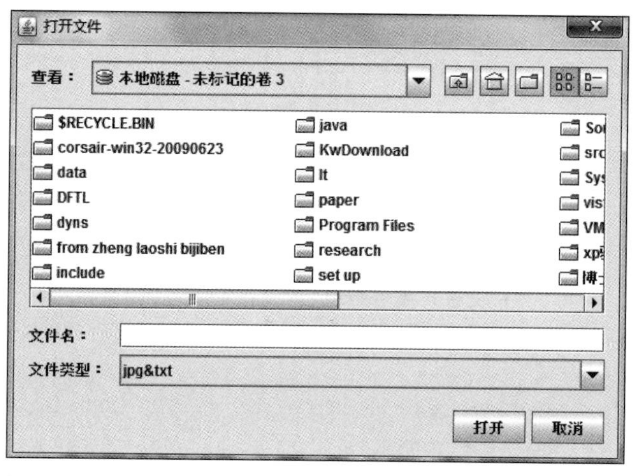


图 7-21　"打开文件"对话框

可见，"文件类型"中的字符串就是传入的 description 对象。当选择"文件类型"为"jpg&txt"时，只能选择 jpg 文件和 txt 文件。

# 第 8 章

# 多线程编程

## 要点导读

本章内容需要与配套的主教材《Java 语言程序设计》(第 3 版)第 8 章配合学习。

主教材第 8 章介绍了多线程。如果要实现一个程序中多段代码同时并发执行，就需要产生多个线程，并指定每个线程上所要运行的程序段，这就是多线程。

在 Java 程序中创建多线程有两种方法：继承 Thread 类和实现 Runnable 接口。从 Thread 类派生一个子类，并创建这个子类的对象，就可以产生一个新的线程。这个子类应该重写 Thread 类的 run() 方法，在 run() 方法中写入需要在新线程中执行的语句段。Runnable 接口也是 Java 多线程机制的一个重要部分，实际上它只有一个 run()方法。实现 Runnable 接口的类的对象可以用来创建线程，这时 start()方法启动此线程就会在此线程上运行 run()方法。

在 Java 技术中，利用对象的“锁”可以实现线程间的互斥操作。每个对象都有一个“锁”与之相连。当线程 A 获得了一个对象的锁后，线程 B 若也想获得该对象的锁，就必须等待线程 A 完成规定的操作并释放出锁后，才能获得该对象的锁并执行线程 B 中的操作。一个对象的锁只有一个，所以利用对一个对象锁的争夺，可以实现不同线程的互斥效果。当一个线程获得锁后，需要该锁的其他线程只能处于等待状态。在编写多线程的程序时，利用这种互斥锁机制，就可以实现不同线程间的互斥操作。

多线程的执行往往需要相互之间的配合。为了更有效地协调不同线程的工作，需要在线程间建立沟通渠道，通过线程间的“对话”来解决线程间的同步问题，而不仅仅是依靠互斥机制。java.lang.Object 类的 wait()、notify()等方法为线程间的通信提供了有效手段。

后台线程(也叫守护线程)通常是为了辅助其他线程而运行的线程，它不妨碍程序终止。

一个线程在任何时刻都处于某种线程状态，主要的状态有：就绪、运行、阻塞。一个新线程的生命从新建一个线程类对象开始，此时线程只是一个空的线程对象，并没有为其分配系统资源。当线程处于这个状态时，只可以对其进行 start 操作。执行了 start()方法后，系统才为新线程创建了资源，并自动调用其 run()函数，线程将处于等待 CPU 资源的就绪状态。获得了 CPU 资源的线程就进入了运行状态。在线程的 run()方法结束时，处于运行状态的线程就进入了死亡状态。处于执行状态的线程由于需要等待某种资源或条件而不得不暂停运行，就会进入阻塞状态。

每个 Java 线程都有一个优先级，其范围都在 Thread.MIN_PRIORITY(常数 1)和 Thread.MAX_PRIORITY(常数 10)之间。默认情况下，每个线程的优先级都设置为

Thread.NORM_PRIORITY(常数 5)。具有较高优先级的线程比优先级较低的线程优先执行。

# 实验 8 线程

## 一、实验目的

(1) 了解线程的概念。

(2) 学会如何通过 Thread 类和 Runnable 接口创建线程,如何实现多线程的资源共享和通信,及如何控制线程的生命。

(3) 掌握线程同步的方法。

(4) 理解线程优先级的概念,以及基于优先级的线程调度。

## 二、实验任务

### 1. 熟练使用 Thread 类和 Runnable 类

分别使用 Thread 和 Runnable 写具有两个同样功能的线程,输出 1~10(要能区分是哪个线程输出的)。启动这两个线程,观看程序输出。将后启动的线程优先级设置为 6,再执行一次,输出结果有区别吗?如果线程优先级设置为 8 呢?输出 1~100、1~1000 时,结果怎样?请简单解释一下程序输出结果。

### 2. 线程的休眠及唤醒

阅读并理解下面的代码,给出一个可能的执行结果,并做简单解释。

```
problem2/CounterThread.java
package problem2;
public class Ex2 {
    public static void main(String[] args) {
        Thread t1 = new Thread(new CounterThread(2, false), "t1");
        Thread t2 = new Thread(new CounterThread(3, false), "t2");
        Thread t3 = new Thread(new CounterThread(4, true), "t3");

        t1.start();
        t2.start();
        try {
            Thread.sleep(2000);
        } catch (InterruptedException e) {
            e.printStackTrace();
        }
        t3.start();
    }
}

problem2/CounterThread.java
```

```
package problem2;
public class CounterThread implements Runnable {
    public static Object lock = new Object();

    int num = 2;

    boolean isNotifier = false;

    public CounterThread(int num, boolean isNotifier) {
        this.num = num;
        this.isNotifier = isNotifier;
    }

    public void run() {
        synchronized (lock) {
            for (int i = 0; i < 5; i++) {
                System.out
                        .println(Thread.currentThread().getName() + " : " + i);
                if (i == num) {
                    if (isNotifier) {
                        lock.notifyAll();
                    } else {
                        try {
                            lock.wait();
                        } catch (InterruptedException e) {
                            e.printStackTrace();
                        }
                    }
                }
            }
        }
    }
}
```

### 3. 生产者、消费者问题

编写生产者(Producer)/消费者(Consumer)程序,要求:

Producer 生产 String,并保存在 ArrayList 中。Consumer 消费(读出并显示)String,并把它从 ArrayList 中删除。生产和消费应有输出显示。

可以指定生产者生产多少个数据,也可以指定消费者消费多少个数据。完成要求数据的生产或消费以后,线程自动终止。应提供相应的构造函数来指定数据个数。

当容器内没有数据时,Consumer 应等待,等容器内有新的数据后再次开始消费。

编写测试程序,测试 1 个 Producer 生产 20 个数据及 2 个 Consumer 各消耗 10 个数据的情形。

#### 4. WareHouse 类实现生产者、消费者问题

将实验任务 3 中的数据容器 ArrayList 替换为本题中的 WareHouse 类。WareHouse 类提供了类似于 ArrayList 的调用接口，size()、getCargo()、addCargo()、removeCargo()方法，分别对应 ArrayList 的 size()、get()、add()、remove()方法。用前面的测试程序测试 1 个 Producer 生产 20 个数据及 2 个 Consumer 各消耗 10 个数据的情形。观察程序输出，输出与实验任务 3 的程序基本一致吗？如果有较大差别，简述原因并改正。

```
import java.util.Vector;

public class WareHouse {
    private static final long DELAY = 100;
    private Vector storage = new Vector();

    public int size() {
        working();
        return storage.size();
    }

    public String getCargo(int i) {
        working();
        return storage.get(i).toString();
    }

    public void addCargo(String cargo) {
        working();
        storage.add(cargo);
    }

    public void removeCargo(int i) {
        working();
        storage.remove(i);
    }

    private void working() {
        try {
            Thread.sleep(DELAY);
        } catch (InterruptedException e) {
            e.printStackTrace();
        }
    }
}
```

## 三、实验步骤

#### 1. 复习理解线程的概念

(1) 复习线程概念，熟悉使用 Thread 类和 Runnable 接口创建线程的方法。

(2) 复习优先级的概念。

(3) 实现实验任务 1 的代码。

**2. 给出实验任务 2 程序的输出**

(1) 复习线程同步、线程通信、线程控制等知识。

(2) 熟悉 synchronized 关键字,wait()、notify()方法。

(3) 在 exp8 工程下创建 problem2 包,完成实验任务 2。

**3. 实现实验任务 3 的代码**

(1) 阅读教材的生产者/消费者问题。

(2) 在 exp8 工程下创建 problem3 包。

(3) 实现 Producer、Consumer 类。

(4) 编写 Test 类,在 main()方法中测试实验任务 3 中要求的场景。

**4. 实现实验任务 4 代码**

(1) 阅读 WareHouse 类的代码。将它放到工程中。

(2) 在 exp8 工程下创建 problem4 包。

(3) 将 problem3 的代码复制到 problem4 包下,注意包定义也要做相应修改。

(4) 在 problem4 包下修改 Producer、Consumer 类。

(5) 执行测试,回答问题。

(6) 修改代码,直到输出和实验任务 3 的程序输出一致的结果。

## 习题解答

1. 进程和线程有何区别? Java 是如何实现多线程的?

解:

每个进程要占有独立内存资源,同一个进程的多个线程是共享内存资源的。另外,线程间的通信也比进程间通信开销小。Java 在语言级提供了对线程的支持,创建多线程有两种方法:继承 Thread 类和实现 Runnable 接口。

2. 简述线程的生命周期,重点注意线程阻塞的几种情况,以及如何重回就绪状态。

解:

一个新线程的生命从新建一个线程类对象开始,此时线程只是一个空的线程对象,并没有为其分配系统资源。当线程处于这个状态时,只可以对其进行 start()操作。进行 start()操作后,线程进入就绪态。处于就绪态的线程被调度后,便进入运行态。处于运行态的线程可能被阻塞在对象的 wait 池或者对象的 lock 池,或者自行调用 sleep()方法,从而进入阻塞态。处于阻塞态的线程在 sleep 的时间到达或者获得锁之后,重新进入就绪态。另外,处于就绪态的线程在执行完 run()方法之后,进入死亡状态。

3. 随便选择两个城市作为预选旅游目标。实现两个独立的线程分别显示 10 次城市名,每次显示后休眠一段随机时间(1000ms 以内),哪个先显示完毕,就决定去哪个城市。分别用 Runnable 接口和 Thread 类实现。

解:

本题的实现较为简单,掌握创建线程对象的两种使用方法即可。

新建 Exe8_3.java 文件。

```
class Thread1 extends Thread {
    @Override
    public void run() {
        for(int i = 0; i < 10; i++) {
            System.out.println("北京");
            try {
                sleep(1000);
            }
            catch(Exception  e) {}
        }
    }
}
class Thread2 implements Runnable {
    @Override
    public void run() {
        for(int i = 0; i < 10; i++) {
            System.out.println("西安");
            try {
                Thread.sleep(1000);
            }
            catch(Exception  e) {}
        }
    }
}
public class Exe8_3 {
    public static void main(String[] args) {
        Thread1 thread1 = new Thread1();
        Thread2 thread2 = new Thread2();
        thread1.start();
        new Thread(thread2).start();
    }
}
```

4. 编写一个多线程程序实现如下功能：线程 A 和线程 B 分别在屏幕上显示信息“…start”后，调用 wait 等待；线程 C 开始后调用 sleep 休眠一段时间，然后调用 notifyAll，使线程 A 和线程 B 继续运行。线程 A 和线程 B 恢复运行后输出信息“…end”后结束，线程 C 在判断线程 B 和线程 C 结束后自己也结束运行。

解：

新建 Exe8_4.java 文件，其内容为：

```
//协助三个线程之间通信的类
class SyncObject {


}
```

```
//线程 A 和 B 的类
class WaitThread extends Thread {
    final SyncObject so;
    String name;
    WaitThread(SyncObject so, String name) {
        this.so = so;
        this.name = name;
    }
    @Override
    public void run() {
        synchronized(so) {     //获得锁
            System.out.println(name + " start");
            try {
                so.wait();
            }
            catch(Exception e) {
                e.printStackTrace();
            }
            so.notifyAll();    //唤醒其他线程
            System.out.println(name + " end");
        }
    }
}
//线程 C 的类
class NotifyThread extends Thread {
    String name;
    final SyncObject so;
    NotifyThread(SyncObject so, String name) {
        this.so = so;
        this.name = name;
    }
    @Override
    public void run() {
        synchronized(so) {
            try {
                sleep(5000);    //sleep 后唤醒其他线程,然后自己 wait
                so.notifyAll();
                so.wait();
            }
            catch(Exception e) {
                e.printStackTrace();
            }
            System.out.println(name + " end");

        }
```

```
    }
}
//测试类,执行 A,B,C 三个线程
public class Exe8_4 {
    public static void main(String[] args) {
        SyncObject so = new SyncObject();
        WaitThread thread1 = new WaitThread(so, "A");
        WaitThread thread2 = new WaitThread(so, "B");
        NotifyThread thread3 = new NotifyThread(so,"C");
        thread1.start();
        thread2.start();
        thread3.start();
    }
}
```

5. 实现一个数据单元,包括学号和姓名两部分。编写两个线程,一个线程往数据单元中写入,另一个线程往外读出。要求每写一次就读出一次。

解:

新建 Exe8_5.java 文件,其内容为:

```
//数据单元类,在其中实现写入和读出的同步
class DataUnit {
    int number;
    String name;
    boolean available;
    DataUnit() {
        this.number = 0;
        this.name = "";
        this.available = false;
    }
    public synchronized void write(int number, String name) {
        if(available) { //如果有内容,则不能写,开始 wait
            try {
                wait();
            }
            catch(Exception e) {
                e.printStackTrace();
            }
        }
        System.out.println("Write" + number + "\t" + name);
        this.number = number;
        this.name = name;
        available = true;
        notify();
    }
```

```
    public synchronized void read() {
        if(!available) {    //如果没有内容,则不能读,开始 wait
            try {
                wait();
            }
            catch(Exception e) {
                e.printStackTrace();
            }
        }
        System.out.println("Read" + number + "\t" + name);
        available = false;
        notify();
    }
}
class DataWriter extends Thread{
    DataUnit du;
    DataWriter(DataUnit du) {
        this.du = du;
    }
    @Override
    public void run() {
        for(int i = 0; i < 10; i++) {          //写入 10 次
            du.write(i, "Name" + i);
        }
    }
}
class DataReader extends Thread {
    DataUnit du;
    DataReader(DataUnit du) {
        this.du = du;
    }
    @Override
    public void run() {
        for(int i = 0; i < 10; i++) {          //读出 10 次
            du.read();
        }
    }
}
public class Exe8_5 {
    public static void main(String[] args) {
        DataUnit du = new DataUnit();
        DataWriter dw = new DataWriter(du);
        DataReader dr = new DataReader(du);
        dr.start();
        dw.start();
    }
}
```

每次运行,程序的输出结果都一样。

```
Write 0         Name0
Read 0          Name0
Write 1         Name1
Read 1          Name1
Write 2         Name2
Read 2          Name2
Write 3         Name3
Read 3          Name3
Write 4         Name4
Read 4          Name4
Write 5         Name5
Read 5          Name5
Write 6         Name6
Read 6          Name6
Write 7         Name7
Read 7          Name7
Write 8         Name8
Read 8          Name8
Write 9         Name9
Read 9          Name9
```

6. 创建两个不同优先级的线程,都从 1 数到 10 000,看看哪个数得快。

解:

新建 Exe8_6.java 文件,其内容为:

```
class CountThread extends Thread {
    int num;
    String name;
    public CountThread(int num, String name) {
        this.num = num;
        this.name = name;
    }
    @Override
    public void run() { //在 run()方法中输出每次数数
        for(int i = 0; i < num; i++) {
            System.out.println(name + ":" + i);
        }
        System.out.println(name + " counts finish");
    }
}
public class exe8_6 {
    public static void main(String[] args) {
        CountThread ct1 = new CountThread(50000, "thread 1");
```

```
        CountThread ct2 = new CountThread(50000, "thread 2");
        ct1.setPriority(2);          //设置第一个线程优先级为 2
        ct2.setPriority(8);          //设置第二个线程优先级为 8
        ct1.start();                 //启动线程
        ct2.start();
    }
}
```

可以看到,结果为 ct2 先数完。读者可以将 run()方法的循环中的输出语句去掉,看看结果如何。此时,可以看到并不是每次都是 ct2 先数完,这是因为 run()方法可以很快执行完,ct1 和 ct2 的差别并不明显。

7. 编写一个程序来说明较高优先级的线程通过调用 sleep 方法,使较低优先级的线程获得运行的机会。

解:

新建 Exe8_7.java 文件,其内容为:

```
package exe8_7;
import java.util.*;
//两个线程共享
class SyncObject {
    public int flag;          //标记位,用于控制线程 2 是否 sleep 以及是否输出
}
class Thread1 extends Thread{
    SyncObject so;
    Thread1 (SyncObject so) {
        this.so = so;
    }
    @Override
    public void run() {
        //线程 1 输出 2500 次,此时线程 2 不 sleep
        Calendar start = Calendar.getInstance();
        for(int i = 0; i < 2500; i++) {
            System.out.println("thread1:" + i);
        }
        Calendar end = Calendar.getInstance();
        long time1 = end.getTimeInMillis() - start.getTimeInMillis();

        so.flag = 1;//设置 flag 为 1,此后线程 2 将 sleep
        start = Calendar.getInstance();
        for(int i = 0; i < 2500; i++) {
            System.out.println("thread1 again:" + i);
        }
        end = Calendar.getInstance();
        long time2 = end.getTimeInMillis() - start.getTimeInMillis();
```

```
            System.out.println("第一次输出,用时:" + time1);
            System.out.println("第二次输出,用时:" + time2);
            so.flag = 2;                    //设置 flag 为 2,此后线程 2 将不输出,便于查看结果
        }
    }
    class Thread2 extends Thread {
        SyncObject so;
        Thread2(SyncObject so) {
            this.so = so;
        }
        @Override
        public void run() {
            for(int i = 0; i < 50000; i++) {
                if(so.flag == 1) {          //若 flag 为 1, 则 sleep
                    try {
                        sleep(1000);
                    }
                    catch(Exception e) {
                        e.printStackTrace();
                    }
                }
                else if(so.flag == 0)       //若 flag 为 0,则线程 2 将不输出,便于查看结果
                    System.out.println("thraed2:" + i);
            }
        }
    }
    public class Exe8_7 {
        public static void main(String[] args) {
            SyncObject so = new SyncObject();
            so.flag = 0;
            Thread1 tr1 = new Thread1(so);
            Thread2 tr2 = new Thread2(so);
            tr1.setPriority(2);
            tr2.setPriority(8);             //设置线程 2 优先级高于线程 1
            tr1.start();
            tr2.start();
        }
    }
```

程序的最后两行输出如下。

```
第一次输出,用时:700
第二次输出,用时:315
```

由结果可以看出,当线程 1 第二次循环输出时,由于线程 2 每循环一次便 sleep,所以线程 1 的第二次输出用时比第一次输出少。

8. 编程实现主线程控制新线程的生命，当主线程运行一段时间后，控制新线程死亡，主线程继续运行一段时间后结束。

解：

新建 Exe8_8.java 文件，其内容为：

```
package exe8_8;
class SonThread extends Thread {
    int flag = 0;
    @Override
    public void run() {
        for (int i = 0; i < 5000; i++) {
            if (flag == 0) {
                System.out.println("i = " + i);
            } else {
                return;
            }
        }
    }
}
public class Exe8_8 {
    public static void main(String[] args) {
        SonThread t = new SonThread();
        t.start();
        for (int j = 0; j < 200; j++) {
            System.out.println("j = " + j);
        }
        t.interrupt();
        t.flag = 1;
        for (int j = 2000; j < 4000; j++) {
            System.out.println("j = " + j);
        }
    }
}
```

从程序的输出结果可以看出，线程 1 被打断后，由于其 flag 被设置为 1，所以从 run()方法中返回，线程 1 结束。

9. 用两个线程模拟存、取货物。一个线程往一对象里放货物(包括品名、价格)，另外一个线程取货物，分别模拟“放一个、取一个”和“放若干个、取若干个”两种情况。

解：

新建 Exe8_9.java 文件，其内容为：

```
package exe8_9;
import java.util.*;
class Goods {
```

```
    Goods(String name, float price) {
        this.name = name;
        this.price = price;
    }
    String name;
    float price;
}
class MyList {
    MyList() {
        goodsList = new ArrayList<Goods>();
    }
    ArrayList<Goods> goodsList;
    //放一个
    public synchronized void put(Goods g) {
        goodsList.add(g);
        notifyAll();
    }
    //放若干个
    public synchronized void put(Goods[] g) {
        goodsList.addAll(java.util.Arrays.asList(g));
        notifyAll();
    }
    //取一个
    public synchronized Goods get() {
        while (goodsList.size() == 0) {
            try {
                wait();
            } catch (Exception e) {
                //e.printStackTrace();
            }
        }
        Goods g = goodsList.remove(0);
        return g;
    }
    //取若干个,个数为 n
    public synchronized Goods[] get(int n) {
        while (goodsList.size() < n) {
            try {
                wait();
            } catch (Exception e) {
                //e.printStackTrace();
            }
        }
        Goods[] g = new Goods[n];
        for (int i = 0; i < n; i++) {
```

```
            g[i] = goodsList.remove(0);
        }
        return g;
    }
}
class MyThread1 extends Thread {
    MyList goodsList;
    MyThread1(MyList goodsList) {
        this.goodsList = goodsList;
    }
    @Override
    public void run() {
        //循环 10 次,每次放一个
        for (int i = 1; i <= 10; i++) {
            goodsList.put(new Goods("thread1:" + i, (float) i / 10));
            System.out.println(i + ":Thread1 put");
        }
    }
}
class MyThread2 extends Thread {
    MyList goodsList;
    MyThread2(MyList goodsList) {
        this.goodsList = goodsList;
    }
    @Override
    public void run() {
        //循环 10 次,每次取一个
        for (int i = 1; i <= 10; i++) {
            goodsList.get();
            System.out.println(i + ":Thread2 get");
        }
    }
}
class MyThread3 extends Thread {
    MyList goodsList;
    MyThread3(MyList goodsList) {
        this.goodsList = goodsList;
    }
    @Override
    public void run() {
        //循环 10 次,每次放两个
        for (int i = 1; i <= 10; i++) {
            Goods[] goods = new Goods[2];
            goods[0] = new Goods("thread3:" + i, (float) i / 20);
            goods[1] = new Goods("thread3:" + i, (float) i / 50);
```

```
                goodsList.put(goods);
                System.out.println(i + ":Thread3 put");
            }
        }
    }
    class MyThread4 extends Thread {
        MyList goodsList;
        MyThread4(MyList goodsList) {
            this.goodsList = goodsList;
        }
        @Override
        public void run() {
            //循环10次,每次取两个
            for (int i = 1; i <= 10; i++) {
                goodsList.get(2);
                System.out.println(i + ":Thread4 get");
            }
        }
    }
    //测试类
    public class Exe8_9 {
        public static void main(String[] args) {
            MyList goodsList = new MyList();
            MyThread1 thread1 = new MyThread1(goodsList);
            MyThread2 thread2 = new MyThread2(goodsList);
            MyThread3 thread3 = new MyThread3(goodsList);
            MyThread4 thread4 = new MyThread4(goodsList);
            thread1.start();
            thread2.start();
            thread3.start();
            thread4.start();
        }
    }
```

程序运行时某一次的结果如下。

```
1:Thread1 put
1:Thread2 get
1:Thread4 get
1:Thread3 put
2:Thread2 get
2:Thread1 put
2:Thread4 get
2:Thread3 put
3:Thread2 get
```

```
3:Thread1 put
3:Thread4 get
3:Thread3 put
4:Thread2 get
4:Thread1 put
4:Thread4 get
4:Thread3 put
5:Thread2 get
6:Thread2 get
7:Thread2 get
5:Thread1 put
5:Thread3 put
8:Thread2 get
6:Thread1 put
5:Thread4 get
6:Thread3 put
9:Thread2 get
7:Thread1 put
6:Thread4 get
7:Thread3 put
10:Thread2 get
8:Thread3 put
8:Thread1 put
7:Thread4 get
9:Thread3 put
8:Thread4 get
9:Thread1 put
9:Thread4 get
10:Thread3 put
10:Thread4 get
10:Thread1 put
```

10. 用两个线程模拟对话，任何一个线程都可以随时收发信息。

解：

新建 Exe8_10.java 文件，其内容为：

```
package exe8_10;
class MyThread extends Thread{
    String name;
    StringBuffer sendString;
    StringBuffer recvString;
    MyThread(String name, StringBuffer sendString, StringBuffer recvString) {
        this.name = name;
        this.sendString = sendString;
        this.recvString = recvString;
```

```
    }
    @Override
    public void run() {
        //循环 100 次,每次循环都 sleep 一段随机时间,然后收发消息
        for(int i = 0; i < 100; i++) {
            int time = (int)(Math.random() * 100);
            try {
                sleep(time);
            }
            catch(Exception e) {
                e.printStackTrace();
            }
            sendMsg();
            recvMsg();
        }
    }
    //发送消息
    public void sendMsg() {
        synchronized(sendString) {
            while(sendString.length() > 0) {
                try {
                    sendString.wait();
                }
                catch(Exception e) {
                    e.printStackTrace();
                }
            }
            double content = Math.random();
            sendString.append(name + content);
            System.out.println(name + " send " + content);
            sendString.notify();
        }
    }
    //接收消息
    public void recvMsg() {
        synchronized(recvString) {
            while(recvString.length() == 0) {
                try {
                    recvString.wait();
                }
                catch(Exception e) {
                    e.printStackTrace();
                }
            }
            System.out.println(name + " recv " + recvString);
```

```
                recvString.delete(0, recvString.length());
                recvString.notify();
            }
        }
    }
    //测试类
    public class Exe8_10 {
        public static void main(String[] args) {
            StringBuffer string1 = new StringBuffer();
            StringBuffer string2 = new StringBuffer();
            MyThread thread1 = new MyThread("A", string1, string2);
            MyThread thread2 = new MyThread("B", string2, string1);
            thread1.start();
            thread2.start();
        }
    }
```

# 第 9 章

# JDBC 编程

## 要点导读

本章内容需要与配套的主教材《Java 语言程序设计》(第 3 版)第 9 章配合学习。

主教材第 9 章介绍了 JDBC 与数据库访问。使用数据库对数据资源进行管理，可以减少数据的冗余度，节省数据的存储空间，实现数据资源的充分共享，为用户提供管理数据的简便手段。数据模型就是数据库的逻辑结构。关系模型是一种数据模型，形象地说，就是二维表结构，也称为关系表。关系数据库就是支持关系模型的数据库。

为了保证关系表中没有重复的记录，可以为关系表定义一个主码。主码又称主键，作用是唯一标识表中的一个记录。

SQL 是关系数据库的标准语言。通过 SQL，可以实现对数据库的各种操作，例如，创建表、插入、删除、修改、查询等。

在 Java 程序中使用 JDBC 连接和访问数据库。在 JDBC 中，可以使用 Class.forName() 方法显式装载驱动程序，使用 DriverManager.getConnection 连接数据库，而 Statement 接口提供了三种执行 SQL 语句的方法：executeQuery()、executeUpdate()和 execute()，用于执行 SQL 语句。

从 Java 6 开始，JDK 自带一个数据库，叫作 Java DB。Java DB 是用 Java 实现的开源数据库管理系统。通过 Java DB，程序员可以省掉安装和配置外部数据库的过程，并能方便地进行数据库编程。

## 实验 9　JDBC 编程

### 一、实验目的

(1) 掌握 JDBC 的概念。

(2) 掌握使用 JDBC 操作数据库。

(3) 掌握简单的 SQL 的使用。

### 二、实验任务

#### 1. 建立一个数据库

建立一个学生及成绩的数据库，在其中建立一个表示学生的表 Student 和表示成绩的表 Score。Student 表的内容包括：姓名、学号、性别，其中学号不能重复。Score 表的内容包括：学号、语文、数学、英语、计算机、政治。

**2. 对数据库进行增、删、改**

实现学生的增加、删除、修改操作。实现学生成绩的增加、删除、修改操作。

注意：当删除学生时，如果已经录入了该学生的成绩，则同时删除该学生的成绩。

## 三、实验步骤

**1. 建立数据库 test**

建立一个 MySQL 数据库命名为 test，在其中用 CREATE TABLE 新建两个表：Student 和 Score。每个表的内容参照实验任务 1 的要求。具体方法参见主教材第 9 章。

**2. 创建新的 Java 项目——exp9**

**3. 新建 Student.java 和 Score.java 文件**

分别实现 Student 类和 Score 类，表示学生和分数。

**4. 新建 StudentOperator 类和 ScoreOperator 类**

实现学生和成绩的操作。

StudentOperator 类的主要方法代码提示如下。

```
//添加一个学生,如果成功,则返回 0;否则返回-1
public static int addStudent(Student s, Statement stmt) throws Exception{
        //首先查找该学号的学生是否存在,如果存在,则不能添加
        int err = 0;
        ...
        return err;
}
//添加一个学生及其成绩
public static int addStudent(Student s, Score score, Statement stmt) throws Exception{

}
//根据学号删除一个学生
public static int removeStudent(String number, Statement stmt) throws Exception{
        //首先删除学生
        ...
        //其次在 Score 表中查找是否有该学生的成绩,如果有,也删除

//根据 s 的学号,修改该学生的信息
public static int updateStudent(Student s, Statement stmt) throws Exception{

}
//显示 Student 表中所有记录
public static void showStudents(Statement stmt) throws Exception{

}
```

ScoreOperator 类的主要方法代码提示如下。

```
//添加成绩
public static int addScore(Score s, Statement stmt) throws Exception{
    //首先查找 s 对应学号的学生是否存在,如果不存在则返回-1
    //如果学生存在,再查找该学号是否已经录入成绩
    //如果已经录入成绩,则返回失败,否则在 Score 表中添加成绩
}
//删除某学号对应的学生成绩
public static int deleteScore(String number, Statement stmt) throws Exception{
    int err = 0;
    String sql = "Delete From Score WHERE Number='" + number + "'";
    err = stmt.executeUpdate(sql);
    return err;
}

public static int updateScore(Score s, Statement stmt) throws Exception {

}
//显示 Score 表中所有记录
public static void showScores(Statement stmt) throws Exception{

}
```

在这些方法中,都通过 JDBC 操作数据库,完成学生和成绩的增加、删除、修改功能。

### 5. 编写测试类 Test.java

检查实现是否正确。测试程序示例如下。

```
public class Test {
    public static void main(String[] args) throws Exception {
        String DBDriver = "com.mysql.jdbc.Driver";
        String connectionStr = "jdbc:mysql://localhost/exp9";
        Connection con = null;
        Statement stmt = null;
        ResultSet rs = null;
        Class.forName(DBDriver);           //加载驱动器
        con = DriverManager.getConnection(connectionStr, "Test", "124");
                                           //连接数据库
        stmt = con.createStatement();      //创建 Statement 对象

        Student s1 = new Student("A", "201", 'M');
        StudentOperator.addStudent(s1, stmt);
        System.out.println("After add s1:");
        StudentOperator.showStudents(stmt);
```

```
        Student s2 = new Student("B", "202", 'F');
        StudentOperator.addStudent(s2, stmt);
        System.out.println("After add s2:");
        StudentOperator.showStudents(stmt);

        StudentOperator.removeStudent(s1.number, stmt);
        System.out.println("After remove s1:");
        StudentOperator.showStudents(stmt);

        s2.name = "C";
        StudentOperator.updateStudent(s2, stmt);
        System.out.println("After update s2:");
        StudentOperator.showStudents(stmt);

        Score score1 = new Score("202", 90, 89, 88, 87, 86);
        ScoreOperator.addScore(score1, stmt);
        System.out.println("After add score1:");
        ScoreOperator.showScores(stmt);

        score1.chinese = 95;
        ScoreOperator.updateScore(score1, stmt);
        System.out.println("After update score1:");
        ScoreOperator.showScores(stmt);

        ScoreOperator.deleteScore(score1.number, stmt);
        System.out.println("After delete score1:");
        ScoreOperator.showScores(stmt);
        stmt.close();       //关闭语句
        con.close();        //关闭连接
    }
}
```

## 习题解答

1. 简述 Java 访问数据库的机制和需要注意的问题。

解:

Java 通过 JDBC(Java DataBase Connectivity)连接和访问数据库,JDBC 是为 Java 语言定义的一个 SQL 调用级的数据库编程接口。

JDBC 使用已有的 SQL 标准,并支持其他数据库连接标准,如与 ODBC 之间的桥接。ODBC 是一个 C 语言实现的访问数据库的 API,对于没有提供 JDBC 驱动的数据库,从 Java 程序调用本地的 C 程序访问数据库会带来一系列安全性、完整性、健壮性等方面的问题,因而通过 JDBC-ODBC 桥来访问没有提供 JDBC 接口的数据库是一个常用的方案。

2. 修改例 9-1，使其能够显示出员工的所有信息，依次为员工编号、员工姓名、部门编号、职务、工资、学历编号。

解：

将例 9-1 中的"while (rs.next())"循环语句替换为如下代码。

```
while (rs.next())      //显示所有记录的信息
{
    System.out.print(rs.getInt("ID") + "  ");
    System.out.print(rs.getString("Name") + "  ");
    System.out.print(rs.getInt("Department") + "  ");
    System.out.print(rs.getString("Occupation") + "  ");
    System.out.print(rs.getInt("Salary")+ "  ");
    System.out.println(rs.getInt("Education"));
}
```

3. 修改习题 2，使程序能显示部门编号对应的部门名称，以及学历编号对应的名称。

解：

主要修改查询的 SQL 语句，将相应代码改成如下代码。

```
rs = stmt.executeQuery("Select t1.*, t2.DepName, t3.EduName From " +
        "Person as t1, Department as t2, Education as t3 " +
        "where t1.Department = t2.DepID and t1.Education = t3.EduID");  //查询表
while (rs.next())      //显示所有记录的信息
{
    System.out.print(rs.getInt("ID") + "  ");
    System.out.print(rs.getString("Name") + "  ");
    System.out.print(rs.getString("DepName") + "  ");
    System.out.print(rs.getString("Occupation") + "  ");
    System.out.print(rs.getInt("Salary")+ "  ");
    System.out.println(rs.getString("EduName"));
}
```

4. 基于上面的程序，增加查错功能：如果程序要删除一条并不存在的员工记录，则给出提示信息，表明不存在该员工，程序继续往下执行；如果程序要添加一条重复编号的员工记录，也给出提示信息，表明该编号已经使用，要求重新输入一个新的编号。

解：

使用 try catch 语句捕获异常，用于处理删除并不存在记录、增加已有编号记录时出现的 SQLException。

插入操作的代码如下。

```
try {
    stmt.executeUpdate("INSERT INTO Person VALUES(9,'林时'," +
            "3,'accountant',2000,4)");    //添加一条记录
}
```

```
catch (SQLException e) {
    System.out.println("编号为 9 的员工已经存在,插入失败");
}
```

删除操作的代码如下。

```
try {
    stmt.executeUpdate("DELETE FROM Person WHERE Name='林时'");
}
catch(Exception e) {
    System.out.println("该用户不存在");
}
```

5. 建立一个基于图形用户界面的 department 表的维护程序，要求从界面输入一个部门号(depId)及一个新的部门名称，程序根据部门号将指定部门更名为新的名字。本题要求使用带参数的 SQL 语句。

解：

新建 Exe9_5.java 文件，其内容为：

```
package exe9_5;
import java.sql.*;
import java.awt.event.*;
import java.awt.*;
import javax.swing.*;
public class Exe9_5 extends JFrame {
    JTextField text1;
    JTextField text2;
    Exe9_5() throws Exception {
        super("更新部门");
        Container contentPane = getContentPane();
        contentPane.setLayout(new BorderLayout());
        JLabel label1 = new JLabel("部门 ID:");
        text1 = new JTextField();
        JLabel label2 = new JLabel("新的部门名称:");
        text2 = new JTextField();
        JPanel panel = new JPanel();
        panel.setLayout(new GridLayout(2, 2));
        panel.add(label1);
        panel.add(text1);
        panel.add(label2);
        panel.add(text2);
        contentPane.add(panel, BorderLayout.CENTER);
        JButton btn = new JButton("更新");
        contentPane.add(btn, BorderLayout.SOUTH);
```

```
        btn.addMouseListener(new MouseAdapter() {
            @Override
            public void mouseClicked(MouseEvent e) {
                int depID = Integer.parseInt(text1.getText());
                String depName = text2.getText();
                int ret = updateDepName(depID, depName);
                //根据不同的返回值,进行不同提示
                if (ret == -2) {
                    JOptionPane.showMessageDialog(null, "数据库连接错误");
                }
                else if (ret == -1) {
                    JOptionPane.showMessageDialog(null, "不存在 ID 为"
                            + depID + "的部门");
                }
                else {
                    JOptionPane.showMessageDialog(null, "修改成功");
                }
            }
        });
        setSize(200, 150);
        setVisible(true);
    }
    public static void main(String[] args) {
        try {
            Exe9_5 application = new Exe9_5();
            application.setDefaultCloseOperation(JFrame.EXIT_ON_CLOSE);
        }
        catch (Exception e) {
            e.printStackTrace();
        }
    }
    public int updateDepName(int depID, String depName) {
        String DBDriver = "sun.jdbc.odbc.JdbcOdbcDriver";
        String connectionStr = "jdbc:odbc:PIMS";
        Connection con = null;
        try {
            Class.forName(DBDriver);      //加载驱动器
            con = DriverManager.getConnection(connectionStr, "Test", "1234");
        }
        catch (Exception e) {
            return -1;
        }
        try {
            //设置两个参数
            String sq = "UPDATE Department SET DepName=? WHERE DepID=? ";
```

```
            PreparedStatement pstmt = con.prepareStatement(sq);
            pstmt.setString(1, depName);        //为第 1 个参数赋值
            pstmt.setInt(2, depID);             //为第 2 个参数赋值
            pstmt.executeUpdate();              //执行 SQL 语句
            pstmt.close();                      //关闭语句
            con.close();                        //关闭连接
        }
        catch (SQLException e) {
            return -2;
        }
        return 0;//更新成功,返回 0
    }
}
```

程序运行后的初始界面如图 9-1 所示。

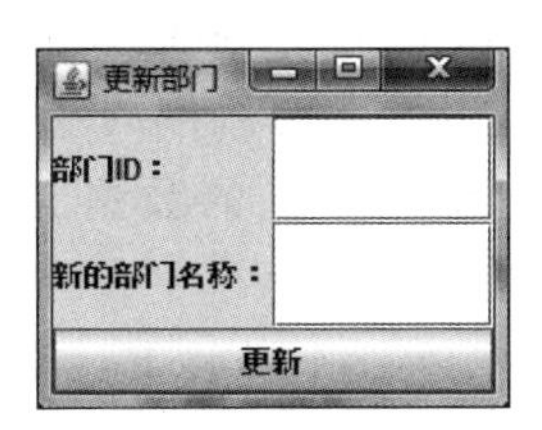


图 9-1　习题 5 程序运行初始界面

6. 一个杂货店老板需要对其货物建立数据库,要求可以输入和查询货物的品名、价格、库存数、厂家电话等,请帮他实现。

解:

建立数据库,设计一个表 Goods,包含如下字段。

ID:自动编号,表示主键。

Name:文本,表示品名。

Price:单精度数字,表示价格。

Stock:长整型数字,表示库存数。

Phone:文本,表示厂家电话。

# 第10章

# Servlet程序设计

## 要点导读

本章内容需要与配套的主教材《Java语言程序设计》(第3版)第10章配合学习。

主教材第10章介绍了Servlet程序设计。所有使用或实现某种Internet服务的程序都必须遵从一个或多个网络协议,IP、TCP、UDP是最根本的三种协议,是所有其他协议的基础。HTTP表示超文本传输协议,它构成了万维网的基础。

统一资源标识符(URI)标识定位网络上的数据文件。通常所说的URL是URI的一种。

面向连接的操作使用TCP协议,在这个模式下一个socket必须在发送数据之前与目的地的socket取得连接。一旦连接建立了,socket就可以使用一个流接口完成打开—读—写—关闭等操作。所有发送的信息都会在另一端以同样的顺序被接收。

UDP是一种无连接的协议,它不能保证数据会被成功地送达,也不保证数据抵达的次序与送出的次序相同。UDP协议可靠性不高,但速度很快。

要运行Servlet就需要Servlet容器,也称为Servlet引擎。Servlet容器是一个编译好的可执行程序,它是Web服务器与Servlet间的媒介,负责将请求翻译成Servlet能够理解的形式传递给Servlet,同时传给Servlet一个对象使之可以返回响应。容器也负责管理Servlet的生命周期。一个Web服务器是能够处理HTTP请求的服务器,它可以提供静态页面、图像等。应用服务器可以支持Servlet和JSP。Servlet容器可以与Web服务器协作提供对Servlet的支持,一些Servlet容器(如Apache Tomcat)自己也可以作为独立的Web服务器运行。

Servlet一般扩展自HttpServlet,并根据浏览器指定的方法(GET或者POST等),重写doGet或者doPost方法。

创建Servlet实例时,它的init方法都会被调用,之后,针对每个客户端的每个请求,都会创建一个线程,该线程调用Servlet实例的Service方法。Service方法根据收到HTTP请求的类型,调用doGet、doPost或者其他方法。最后,如果需要卸载某个Servlet,服务器调用Servlet的destroy方法。

Servlet API提供了两种可以跟踪客户端状态的方法,分别是使用Cookie和使用Session。

# 实验 10　Servlet 程序设计

## 一、实验目的

(1) 掌握 Tomcat 的安装方法和在 Tomcat 部署 Servlet 的方法。

(2) 了解 Servlet 的原理。

(3) 掌握简单的 Servlet 的程序设计方法。

## 二、实验任务

(1) 使用 Servlet 实现实验 9 建立的数据库中的学生信息显示。

在实验 9 的基础上,通过查找数据库,使用 Servlet 实现学生信息的显示。

(2) 附加题:使用 Servlet 实现实验 9 的学生成绩显示。

在实验 9 的基础上,通过查找数据库,使用 Servlet 实现所有学习成绩的显示。

## 三、实验步骤

### 1. 复习原理和概念

复习 Servlet 的原理和概念,掌握 HTML 文件的生成方法。

### 2. 新建项目

在 IntelliJ 中新建项目,其中 Archetype 选择 maven-archetype-webapp。

### 3. 添加目录

在 File→Project Structure→Project Settings→Libraries 下添加"javax-servlet-javax.servlet-api/x.x.x"目录和"apache-tomcat-x.x.xx/lib"目录。

### 4. 在项目工程文件 pom.xml 中添加 servlet 依赖

```
<dependency>
<groupId>javax.servlet</groupId>
<artifactId>javax.servlet-api</artifactId>
<version>4.0.1</version>
<scope>provided</scope>
</dependency>
```

### 5. 新建一个 Servlet 文件 Exp10.java 显示所有学生信息

参考主教材例题 10-7 中使用 Servlet 生成表格的方法,以表格的方式显示学生信息,表格每一行表示一个学生。

### 6. 在 WEB-INF 下 web.xml 文件中添加 url 映射

```
<servlet-mapping>
    <servlet-name>Exp10</servlet-name>
  <url-pattern>/exp10</url-pattern>
</servlet-mapping>
```

7. 运行观察

运行并访问 localhost:xxxx(端口号)/exp10/exp10,观察是否正确显示学生信息。

8. 参考上述步骤,实现学生成绩的显示(附加,选做)

## 习题解答

1. 比较 TCP 与 UDP 两个协议的差别。

解:

TCP 提供可靠的传输,定义了网络上程序到程序的数据传输格式和规则,提供了 IP 数据包的传输确认、丢失数据包的重新请求、将收到的数据包按照它们的发送次序重新装配的机制。TCP 协议是面向连接的协议,在开始数据传输之前,必须先建立明确的连接。和 TCP 相比,UDP 是不可靠的,不保证数据一定传输到终点,也不提供重新排列次序或重新请求功能。与 TCP 的有连接相比,UDP 协议是一种无连接协议。但是,UDP 比 TCP 具有更好的传输效率。

2. 什么是 URI,其作用是什么?

解:

URI 是统一资源标识符的简称,可以用来标识定位网络上的数据文件。

3. 基于套接字的通信中,用 Java 建立简单服务器程序的步骤是什么?

解:

用 Java 建立简单的服务器程序需要以下 5 个步骤。

(1) 调用 ServerSocket 类的构造方法创建 ServerSocket 对象。

(2) 服务器通过 ServerSocket 的 accept 方法监听客户连接。有一个客户连接时,将产生并返回一个 socket。

(3) 获取 InputStream 和 OutputStream 对象,服务器通过接收发送字节与客户进行通信。

(4) 客户和服务器通过 OutputStream 和 InputStream 对象进行通信。

(5) 通信传输完毕,服务器通过调用流和套接字的 close 方法关闭连接。

4. 与传统 CGI(通用网关接口)相比,Servlet 的优点是什么?

解:

与传统的 CGI 相比,Servlet 具有高效率、应用方便、功能强大、便携性好、安全性高和成本低等优点。

5. 练习配置 Apache Tomcat 服务器软件。

解:

见主教材 10.2.2 节。

6. 编写 Servlet 程序,练习 Cookie 的创建、发送和读取。

解:

见主教材 10.5.1 节和例 10-8。

7. 编写 Servlet 程序,练习获得、存储和销毁 Session。

解：

见主教材10.5.2节和例10-7。

8. 编写一个用于查询的动态Web应用程序。该程序从一个由Product表构成的数据库中获取信息，以创建动态的Web页面。Product应该有4个域：productID(关键字)、表示名称的productName，表示价格的productPrice和表示产品简介introduction。调用Servlet时，应该根据查询内容从数据库读取数据，然后动态地创建一个包含名称、价格和简介的Web网页。

解：

在MySQL中建立名为“exe10_8”的数据库，在其中建立表Product。

新建工程Exp10，新建Exe10_8.java文件，其内容为：

```
import java.io.*;
import javax.servlet.*;
import javax.servlet.http.*;
import java.util.*;
import java.sql.*;
class Product {
    int productID;
    String productName;
    float productPrice;
    String introduction;
    //构造方法
    Product(int productID, String productName, float productPrice,
            String introduction) {
        this.productID =productID;
        this.productName =productName;
        this.productPrice =productPrice;
        this.introduction =introduction;
    }
}
public class Exe10_8 extends HttpServlet {
    public ArrayList getProducts() {
        String DBDriver ="com.mysql.jdbc.Driver";
        String connectionStr ="jdbc:mysql://localhost/exe10_8";
        Connection con =null;
        Statement stmt =null;
        ResultSet rs =null;
        ArrayList<Product>al =new ArrayList<Product>();
        try {
            Class.forName(DBDriver);        //加载驱动器
            //连接数据库
            con =DriverManager.getConnection(connectionStr, "Test", "1234");
            stmt =con.createStatement();
```

```
            //查询 Product 表
            rs =stmt.executeQuery("Select * From Product");
            while (rs.next()) {
                // 获取表的内容,构造 Product 对象并添加到 al 中
                int productID =rs.getInt("productID");
                String name =rs.getString("productName");
                float price =rs.getFloat("productPrice");
                String intro =rs.getString("inctroduction");
                Product p =new Product(productID, name, price, intro);
                System.out.println(p);
                al.add(p);
            }
            stmt.close();          //关闭语句
            con.close();           //关闭连接
        }
        catch (Exception e) {
            e.printStackTrace();
        }
        return al;
    }
    @Override
    public void doGet(HttpServletRequest req, HttpServletResponse res)
            throws ServletException, IOException {
        ArrayList<Product>al =getProducts();
        res.setContentType("text/html");
        PrintWriter out =res.getWriter();
        out.println("<HTML>");
        out.println("<HEAD><TITLE>Products Listing</TITLE></HEAD>");
        out.println("<BODY>");
        out.println("<B>Products Listing </B>");
        out.println("      <table border=\"1\">\r\n");
        out.println("          <thead>\r\n");
        out.println("              <tr>\r\n");
        out.println("                  <th>Name</th>\r\n");
        out.println("                  <th>Price</th>\r\n");
        out.println("                  <th>Introduction</th>\r\n");
        out.println("              </tr>\r\n");
        out.println("          </thead>\r\n");
        out.println("          <tbody>\r\n");
        //生成表格内容
        for(int i =0; i <al.size(); i+ + ) {
            Product p =al.get(i);
            out.println("              <tr>\r\n");
            out.println("                  <td>" +
                    p.productName +  "</td>\r\n");
```

```
            out.println("               <td>" +
                    p.productPrice + "</td>\r\n");
            out.println("               <td>" +
                    p.introduction + "</td>\r\n");
            out.println("             </tr>\r\n");
        }
        out.println("          </tbody>\r\n");
        out.println("       </table>\r\n");
        out.println("</BODY></HTML>");
    }
}
```

在 web.xml 文件中添加 servlet 映射：

```
<servlet>
  <servlet-name>Exp10</servlet-name>
  <servlet-class>Exp10</servlet-class>
</servlet>
<servlet>
  <servlet-name>Exp10_Score</servlet-name>
  <servlet-class>Exp10_Score</servlet-class>
</servlet>
<servlet-mapping>
  <servlet-name>Exp10_Score</servlet-name>
  <url-pattern>/exp10_score</url-pattern>
</servlet-mapping>
<servlet-mapping>
    <servlet-name>Exp10</servlet-name>
  <url-pattern>/exp10</url-pattern>
</servlet-mapping>
```

编译运行，即可通过 localhost:8080/Exp10/exp10 和 localhost:8080/Exp10/exp10_score 看到 servlet 输出结果。

9. 编写一个由 Servlet 和多个 Web 网页组成的 Web 应用程序。用户看到的第一个文档为 index.html，该文档包含一系列的超链接。每个超链接都由包含 page 参数的 get 请求来调用 Servlet。Servlet 获取 page 参数并将请求重定向到合适的文档。

解：

新建 index.html 文件，其内容为：

```
<! DOCTYPE HTML PUBLIC "-//W3C//DTD HTML 4.01 Transitional//EN">
<html>
  <head>
    <title></title>
    <meta http-equiv="Content-Type" content="text/html; charset=GBK">
```

```
  </head>
  <body>
      <a href="http://localhost:8080/Chap10/exe10_9? value=1">1</a>
      <hr>
      <a href="http://localhost:8080/Chap10/exe10_9? value=2">2</a>
      <hr>
      <a href="http://localhost:8080/Chap10/exe10_9? value=a">a</a>
  </body>
</html>
```

新建 Exe10_9.java 文件,其内容为:

```
import java.io.IOException;
import java.io.PrintWriter;
import javax.servlet.*;
import javax.servlet.http.*;
public class Exe10_9 extends HttpServlet {
    @Override
    protected void doGet(HttpServletRequest request, HttpServletResponse res)
    throws ServletException, IOException {
        //获得 value 的值
        String value =request.getParameter( "value");
        res.setContentType("text/html");
        PrintWriter out =res.getWriter();
        out.println("<HTML>");
        out.println("<HEAD><TITLE>Hello World</TITLE></HEAD>");
        out.println("<BODY>");
        out.println("<BIG>The value is " +  value +  "</BIG>");
        out.println("</BODY></HTML>");
    }
}
```

对两个文件部署后,单击 index.html 中的不同链接,调用 Exe10_9 返回不同的内容。

# 第11章

# JSP程序设计

## 要点导读

本章内容需要与配套的主教材《Java语言程序设计》(第3版)第11章配合学习。

主教材第11章介绍了JSP程序设计。JSP由静态HTML、专有的JSP标签和Java代码组成。JSP文件需要被转换成Servlet,Servlet在编译后,载入服务器内存中,初始化并执行。

除了标准的HTML以外,JSP主要包括三类组件：脚本元素、指令标签、动作标签。其中,脚本元素可以向JSP文件产生的Servlet文件中插入代码;指令标签将影响由JSP页产生的Servlet的总体结构;动作标签是一种特殊的标签,它影响JSP运行时的功能。

JSP的内置对象有out、request、response、session、application等,这些对象是JSP技术非常重要的组成部分。

为了便于Web开发人员实现内容和功能的分离,可以使用JavaBean或者自定义标签。JavaBean使用一组相当简单而又标准的设计和命名约定,因而调用它们的应用程序无须理解其内部工作原理,就可以很容易地使用JavaBean的方法。JavaBean类的一个实例叫作一个bean。自定义的标签需要3部分：实现标签行为的标签处理类、将XML元素名称映射到标签实现上的标签库描述文件,以及使用标签的JSP程序。

通常可以将一个Web应用程序的结构分为3部分：显示层、业务逻辑层、控制层。其中,显示层包括前端的HTML和Applet等,主要用作用户的操作接口,负责让用户输入数据以及显示数据处理后的结果;业务逻辑层负责数据处理、连接数据库、产生数据等;控制层控制整个网站的流程。这3部分分别对应模型-视图-控制器(Model-View-Controller,MVC)架构的视图、模型、控制器。

Web服务是基于XML和HTTPS的一种服务,它向外界提供一个能够通过Web进行调用的API。

## 实验11 JSP程序设计

### 一、实验目的

(1) 掌握在Tomcat中部署JSP的方法。

(2) 了解JSP的原理。

(3) 掌握简单的 JSP 程序设计。

## 二、实验任务

在实验 10 的基础上,使用 JSP 实现学生信息和成绩信息的显示。

## 三、实验步骤

(1) 复习 JSP 的原理和概念。

(2) 创建 IntelliJ Marven webapp 工程,名为 exp11,添加 Servlet、MySQL 等依赖。

(3) 创建 exp11 包,包含 Score、Student、Exp11 三个类,实现对学生和成绩的查询。

(4) 参考例 11-9,创建 Exp11.jsp,借助(import)exp11 包,实现所有学生和成绩信息的显示。

(5) 用 Tomcat 运行服务器,访问 localhost:8080/exp11/Exp11.jsp,观察是否正确输出学生信息和成绩信息。

# 习题解答

1. 简述 JSP 的运行机制,以及 JSP 在 Tomcat 中的部署,并与 Servlet 的运行机制及部署进行对比。

解:

JSP 由静态 HTML、专用的 JSP 标签和 Java 代码组成。实际上,JSP 文档在后台被自动转换成 Servlet。JSP 在 Tomcat 中的部署见主教材 11.1.1 节。

2. 编写一个 JSP 文件,使每次刷新后都可以显示访问次数。

解:

新建 Exe11_2.jsp 文件,其内容为:

```
<%@page contentType="text/html" pageEncoding="GBK"%>
<%
int n=0;
String counter=(String)application.getAttribute("counter");
if(counter!=null)
   n=Integer.parseInt(counter);
   ++n;
counter=String.valueOf(n);
application.setAttribute("counter",counter);
%>
<html>
    <head>
        <meta http-equiv="Content-Type" content="text/html; charset=GBK">
        <title>JSP Page</title>
    </head>
    <body>
        <h1>访问次数为:<%= n %></h1>
```

```
    </body>
</html>
```

3. 修改习题 2 的程序，使访问次数存在硬盘的文件中，以便服务器重启后可以在已有的访问次数基础上继续增加。

解：

新建 Exe11_3.jsp 文件，其内容为：

```
<%@page contentType="text/html" pageEncoding="GBK"%>
<%
int n=0;
//通过文件获得已经访问的次数
java.io.File f = new java.io.File("D:\\counter.txt");
if (!f.exists()) {
    try {
        f.createNewFile();
    } catch (java.io.IOException e) {
        e.printStackTrace();
    }
}
java.io.FileReader fr = new java.io.FileReader(f);
java.io.BufferedReader br = new java.io.BufferedReader(fr);
String counter = br.readLine();
if(counter!=null)
    n=Integer.parseInt(counter);
++n;
br.close();
//将新的访问次数写入文件
java.io.FileWriter fw = new java.io.FileWriter(f);
counter=String.valueOf(n);
fw.write(counter);
fw.close();
%>
<html>
    <head>
        <meta http-equiv="Content-Type" content="text/html; charset=GBK">
        <title>JSP Page</title>
    </head>
    <body>
        <h1>访问次数为:<%= n %></h1>
    </body>
</html>
```

4. 修改 EmployeeBean，使例 11-4 输入中文也能正常显示。

解：

在类 EmployeeBean 中，增加方法 encodeChinese()，如下。

```
String encodeChinese(String s) {
    String retString = new String();
    try {
        //通过 String 类的构造方法将指定的字符串转换为 gb2312 编码
        retString = new String(s.getBytes("ISO-8859-1"), "gb2312");
    }
    catch (Exception e) {
        retString = "";
        e.printStackTrace();
    }
    return retString;
}
```

再在 getName()方法、getOccupation()方法中，返回转换后的字符串，如下。

```
public String getName() {
    return encodeChinese(name);
}
public String getOccupation() {
    return encodeChinese(occupation);
}
```

此时，可以提交中文，图 11-1 所示就是一个例子。

姓名：
性别： 男 女
编号：
职务：
工资：
提交

名字是：姜尚

编号是：2009301457

是否男性？true

工资是：5000元

职务是：丞相

图 11-1　习题 4 程序运行界面

5. 建立一个计算圆面积和周长的 JavaBean，在 JSP 文件中使用这个 JavaBean 来计算给定半径的圆周长和面积。

解：

编写一个计算周长和面积的 JavaBean，名为 CircleBean.java，如下。

```
package myBeans;
public class CircleBean {
    private double radius = 0.0f;
    public double getRadius() {
        return radius;
    }
    public void setRadius(double radius) {
        this.radius = radius;
    }
    public double getCircumference () {
        return Math.PI * 2 * radius;
    }
    public double getArea() {
        return Math.PI * radius * radius;
    }
}
```

新建 Exe11_5.jsp 文件，其内容为：

```
<%@ page contentType="text/html; charset=gb2312"  %>
<html>
<head>
<title>使用 JavaBean</title>
</head>
<body>
    <form name="form1" method="post" action="">
    <br>半径: <input type="text" name="radius">
    <br><input type="submit" name="Submit" value="提交">
    </form>
    <jsp:useBean id="circle" class="myBeans.CircleBean" />
    <jsp:setProperty name="circle" property="radius" param="radius"  />
    <P>半径为<jsp:getProperty name="circle" property="radius" /> 时: </P>
    <P>周长为: <jsp:getProperty name="circle" property="circumference" /></P>
    <p>面积为: <jsp:getProperty name="circle" property="area" /></p>
</body>
</html>
```

Exe11_5.jsp 运行效果如图 11-2 所示。

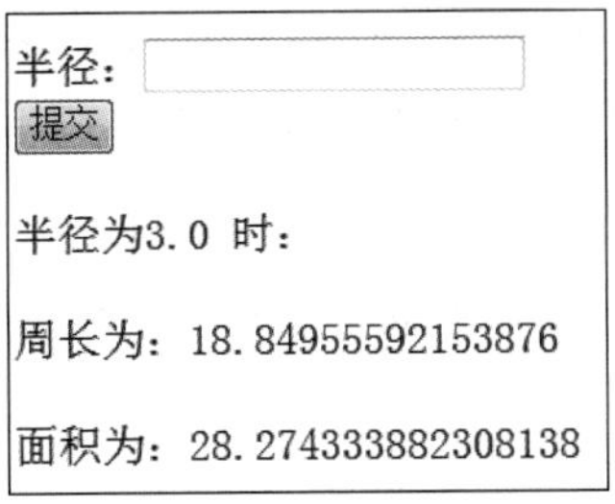

半径：
提交

半径为3.0 时：

周长为：18.84955592153876

面积为：28.274333882308138

图 11-2　习题 5 程序运行界面

6. 建立一个留言信息的 JavaBean，包括留言作者、标题、内容，在 JSP 文件中令该 JavaBean 的应用范围是 application，以便实现多客户多条留言信息的总体显示。

解：

新建 Exe11_6.jsp 文件，其内容为：

```
<%@ page contentType="text/html; charset=gb2312"  %>
<%@page import="java.util.ArrayList" %>
<%@page import="myBeans.Message"%>
<html>
<head>
<title>留言板</title>
</head>
<body>
    <%-- 设置 scope 为 application --%>
    <jsp:useBean id="exe11_6" class="myBeans.Exe11_6" scope="application"/>
    <%
    String author = (String)request.getParameter("author");
    String title = (String)request.getParameter("title");
    String content = (String)request.getParameter("content");
    if(author != null && title != null && content != null) {
        if(author != "" && title != "" && content != "") {
            Message m = new Message(author, title, content);
            exe11_6.getMessageList().add(m);
        }
    }
%>
    <form name="form1" method="post" action="Exe11_6.jsp">
    <br>作者: <input type="text" name="author">
    <br>标题: <input type="text" name="title">
    <br>内容: <input type="textarea" style="width: 300px; height:100px" name=
"content">
    <br><input type="submit" name="Submit" value="提交">
    </form>
    <P>留言列表: </P>
    <table>
<%
      for(int i=0;i<exe11_6.getMessageList().size();i++) {
        Message message=(Message)exe11_6.getMessageList().get(i);
%>
  <tr>
    <th>作者: </th><th><%= message.author%></th>
  </tr>
  <tr>
    <th>标题: </th><th><%= message.title%></th>
```

```
    </tr>
    <tr>
      <th>内容：</th>
      <th><%= message.content%></th>
    </tr>
<%
      }//for 循环结束
%>
      </table>
</body>
</html>
```

在包 myBeans 下建立 Message.java 文件，定义类 Message，表示一条留言，其内容为：

```
package myBeans;
public class Message {
    public String author = "";
    public String title = "";
    public String content = "";
    public Message() {
        this("", "", "");
    }
    public Message(String author, String title, String content) {
        this.author = author;
        this.title = title;
        this.content = content;
    }
}
```

在包 myBeans 下建立 Exe11_6.java 文件，其内容为：

```
package myBeans;
import java.util.*;
public class Exe11_6{
    ArrayList<Message> messageList = null;
    public Exe11_6() {
        messageList = new ArrayList<Message>();
    }
    public ArrayList<Message> getMessageList() {
        return this.messageList;
    }
}
```

浏览 Exe11_6.jsp 时的界面如图 11-3 所示。

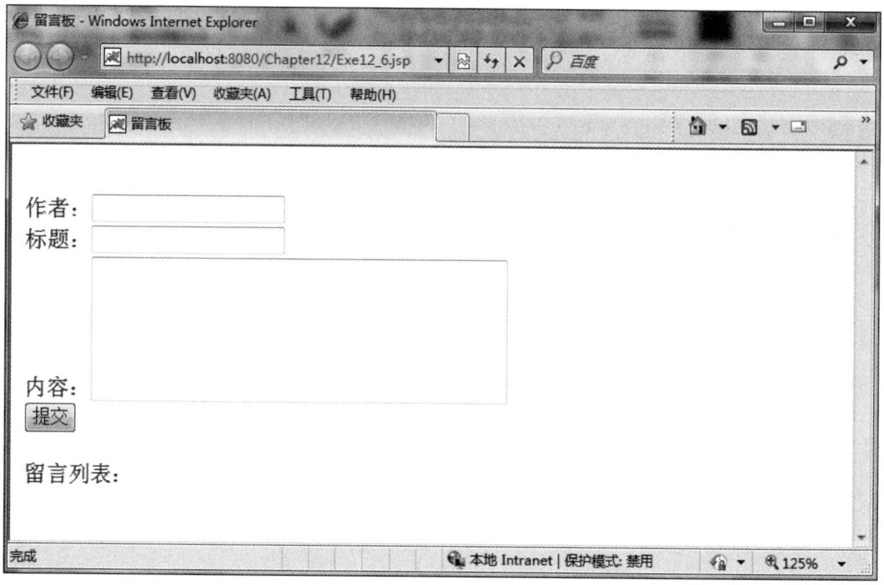

图 11-3　习题 6 程序运行界面

7. 通过在 JSP 程序中使用标签，实现显示某个自然数范围内所有质数的功能。

解：

新建 Exe11_7.jsp 文件，其内容为：

```
<%@page contentType="text/html" pageEncoding="GBK"%>
<html>
    <head>
        <meta http-equiv="Content-Type" content="text/html; charset=GBK">
        <%-- 标签库描述文件的路径及 prefix --%>
        <%@ taglib uri="/WEB-INF/tlds/number" prefix="numberTag" %>
        <title>质数</title>
    </head>
    <body>
<%
            String s = (String)request.getParameter("number");
        %>
        <h1>请输入自然数</h1>
        <form name="form1" method="post" action="">
            <h1><input type="text" name="number"></h1>
            <h1><input type="submit" name="Submit" value="提交"></h1>
        </form>
        <%-- 参数的值可以使用 JSP 表达式,传递输入的整数到标签 --%>
        <H1><numberTag:number num="<%= s %>"/></H1>
    </body>
</html>
```

建立标签库描述文件 number.tld，其内容为：

```
<?xml version="1.0" encoding="UTF-8"? >
<taglib version="2.0" xmlns="http://java.sun.com/xml/ns/j2ee"
xmlns:xsi="http://www.w3.org/2001/XMLSchema-instance"
xsi:schemaLocation="http://java.sun.com/xml/ns/javaee web-jsptaglibrary_2_0.xsd">
  <tlib-version>1.0</tlib-version>
  <short-name>number</short-name>
  <uri>/WEB-INF/tlds/number</uri>
  <tag>
    <name>number</name>
    <tag-class>tags.Number</tag-class>
    <body-content>empty</body-content>
    <description>Get the prime number</description>
    <attribute>
      <name>num</name>
      <required>false</required>
      <!-- 设置 rtexprvalue 为 true,表示可以使用 JSP 表达式 -->
      <rtexprvalue>true</rtexprvalue>
    </attribute>
  </tag>
</taglib>
```

在 tags 包中建立 Number.java 文件，定义 number.tld 文件中关联的 tags.Number 类，其内容为：

```
package tags;
import javax.servlet.jsp.*;
import javax.servlet.jsp.tagext.*;
import java.io.*;
public class Number extends TagSupport {
    protected int num = -1;
    @Override
    public int doStartTag() {
        if(num >= 2)
            printPrimes();
        return (SKIP_BODY);
    }
    public void setNum(String num) {
        if(num == null)
            return;
        try {
            this.num = Integer.parseInt(num);
        }
        catch (NumberFormatException nfe) {
            this.num = 100;
```

```
        }
    }
    //打印质数
    public void printPrimes() {
        try {
            JspWriter out = pageContext.getOut();
            boolean[] isPrime = new boolean[num + 1];
            for(int i = 0; i < num+1; i ++) {
                isPrime[i] = true;
            }
            for(int i = 2; i <=num / 2; i++) {
                for(int j = 2; j <= num / i; j++)
                    isPrime[i * j] = false;
            }
            out.print("小于"+ num +"的质数有: ");
            for(int i = 2; i<num+1; i++) {
                if(isPrime[i])
                    out.println(i);
            }


        }
        catch (IOException ioe) {
            System.out.println("Error in ExampleTag:" + ioe);
        }
    }
}
```

程序运行结果如图 11-4 所示。

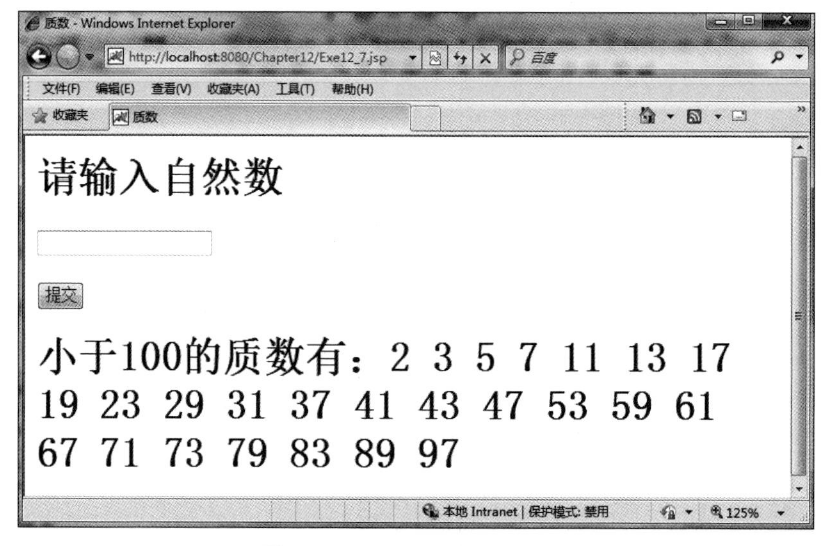


图 11-4　习题 7 程序运行界面

8. 为主教材例 11-9 增加功能，使其还可以从购物车中删除某本书。

解：

将 Checkout.jsp 文件中原有的代码。

```
<table>
…
</table>
```

替换为：

```
<table>
<tr align="center" valign="middle" bgcolor="#CCCCCC">
    <th width="180" scope="col">书名</th>
    <th width="131" scope="col">作者</th>
    <th width="122" scope="col">出版社</th>
    <th width="85" scope="col">价格</th>
    <th width="65" scope="col">数量</th>
    <th width="65" scope="col">操作</th>
</tr>
<%
    Vector buyList=(Vector)session.getAttribute("shoppingcart");
    String str=request.getParameter("id");
    if(str==null) {
    }
    else {
        int r=Integer.parseInt(str);
        String amountString = (String)session.getAttribute("amount");
        float amount = Float.parseFloat(amountString);
        amount -= ((BookBean)buyList.get(r)).getPrice() *
                ((BookBean)buyList.get(r)).getQuantity();
        buyList.remove(r);
        session.setAttribute("amount", "" + amount);//修改总金额
    }
%>
  <form name="form1" method="post" action="Checkout.jsp">
<%
    for (int i = 0; i < buyList.size(); ++i) {
        BookBean aBook=(BookBean)buyList.elementAt(i);
%>
    <tr>
        <th width="180" scope="col"><%= aBook.getName() %></th>
        <th width="131" scope="col"><%= aBook.getAuthor() %></th>
        <th width="122" scope="col"><%= aBook.getPublisher() %></th>
        <th width="85" scope="col"><%= aBook.getPrice() %></th>
        <th width="65" scope="col"><%= aBook.getQuantity() %></th>
```

```
        <th><input type="submit" name="submit" value="删除"/>
            <input type="hidden" name="id" value=<%= i %> />
        </th>
    </tr>
<%
}
%>
</form>
</table>
```

可以在显示结账信息时删除已选图书,如图 11-5 所示。

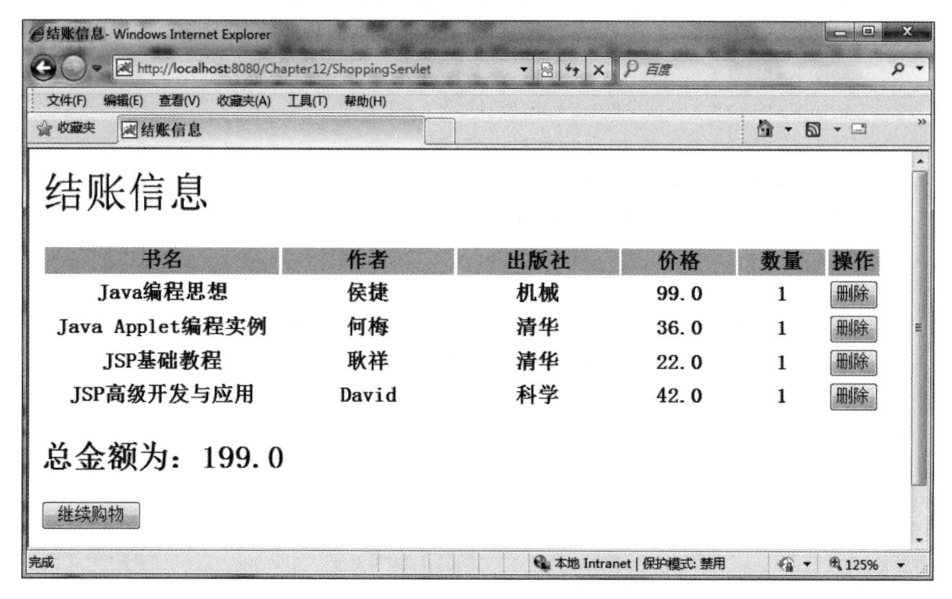


图 11-5 习题 8 程序运行界面

# 第 12 章

## Java 工程化开发概述

### 要点导读

本章内容需要与配套的主教材《Java 语言程序设计》(第 3 版)第 12 章配合学习。

主教材第 12 章对前面章节中学习到的 Java 知识和特性进行应用实践,并在这个过程中体会软件工程的实践思想与实践理念。

核心的软件开发流程包括:需求分析、系统设计、编程实现、测试、试运行、正式运行、系统升级维护、用户服务。开发人员为避免和用户对需求的认识产生巨大偏差,一方面要站在用户的角度深入思考,另一方面要积极与用户沟通。

选择合适的开发工具,建立一套稳定、高效的开发环境,对软件项目的顺利进行至关重要。本章推荐的 Java 环境工具包括:版本控制工具 Git、项目构建工具 Maven 和集成开发环境 IntelliJ IDEA,它们均能支持 Windows、macOS、Linux 等主流操作系统。

框架一般指系统的可重用部分,通常表现为一组抽象的组件以及组件之间的交互方法。在团队内使用统一的开发框架与开发规范,能够降低沟通成本,提高项目可维护性,有利于项目的知识管理与知识积累。

开发系统业务功能,首先要对模块的主要功能进行设计,基于功能设计定义出核心业务服务接口,然后分别开发数据存取程序和展现程序。

程序开发完成后,需要对程序进行完备的测试,以确保程序正确、安全运行,确保软件实现的效果符合预期需求。按照测试的阶段划分,一般可以将测试活动分为单元测试、集成测试、系统测试、验收测试和回归测试五个阶段。

### 实验 12　Java 工程化开发概述

#### 一、实验目的

(1) 掌握使用 Git 进行版本控制的方法。

(2) 掌握使用 Maven 进行项目构建的方法。

#### 二、实验任务

(1) 使用 Git 构建仓库和管理代码。

(2) 使用 Maven 构建项目。

## 三、实验步骤

(1) 使用 git init 命令新建一个 Git 仓库。

(2) 编写一个简单的 HelloWorld.java 代码。

(3) 使用 git add 和 git commit 等命令将代码提交。

(4) 在项目的根目录下新建 pom.xml 文件,配置项目信息,其中需要包含如下内容。

```
    ...
<properties>
    <!-- <project.build.sourceEncoding>UTF-8</project.build.sourceEncoding> -->
    <maven.compiler.source>13</maven.compiler.source>
    <maven.compiler.target>13</maven.compiler.target>
    <maven.compiler.release>13</maven.compiler.release>
</properties>
...
<build>
    <plugins>
        <plugin>
            <groupId>org.apache.maven.plugins</groupId>
            <artifactId>maven-compiler-plugin</artifactId>
            <version>3.9.0</version>
            <configuration>
                <release>13</release>
            </configuration>
        </plugin>
    </plugins>
</build>
...
```

(5) 将源代码放到 src\main\java 目录下。

(6) 在项目的根目录下运行 mvn compile 命令编译项目。

(7) 在项目的根目录下运行 mvn exec:java -Dexec.mainClass=HelloWorld 命令运行项目。

# 习题解答

1. 主教材 12.1 节描述的"院系足球比赛管理系统"需求有哪些不够明确的地方？如果你有半个小时的时间与学生工作组负责赛事组织的老师进行一对一沟通,你会提出哪些问题,从而对系统的需求做进一步的确认,以便顺利开展后续开发工作？

解:

不明确的地方有：参赛队伍都包括哪些信息、分组抽签的规则、积分规则等。

我会提出不限于以下问题：参赛队伍都包括哪些信息？是否需要支持批量导入参赛队伍？积分规则如何(胜平负各积多少分、平分情况如何分名次)？

2. 扩充 FixtureService.java 接口，补充定义以下接口方法：

(1) 查询进球最多的 $N$ 名球员；

(2) 按比赛进球数由多到少排序列出全部比赛；

(3) 查询全部结果为平局的比赛。

解：

修改 FixtureService.java 文件，加入以下接口方法定义。

```
List<Player> nMostGoalsPlayers(int n);
List<Fixture> listByGoals();
List<Fixture> queryTie ();
```

3. 在主教材 12.4 节中定义的 FixtureService.java 接口可能存在哪些性能问题？列出这些问题，并对接口进行相应的修改。

解：

在查询比赛时，返回的是整个 Fixture 对象，包括比赛的所有信息，但用户想查看的信息可能只占一小部分。所以可以定义一些新的接口，支持用户指定查询哪些信息，例如：

```
List<Integer> queryScoresByTeam(string team);      //仅查询比分
List<String> queryOpponentByTeam(string team);     //仅查询对手
```

4. 如果将比赛模块的数据模型设计应用到真实的环境中，可能会产生哪些问题？应该如何修改完善？

解：

没有记录各个球员属于哪个队的信息，或各个队有哪些球员的信息。这就有可能导致录入比赛事件时，若输入的球队与队员不对应，系统无法报错提醒。应该在球员的定义中记录所属球队，在比赛事件的定义中加入限制，检查球员与球队的对应关系。

5. 修改主教材的例 12-9，实现以下测试方法。注意体会如何通过 Mockito 对 repository 的行为进行模拟。

(1) void testListByDateAndTime()。

(2) void testQueryByField()。

(3) void testQueryByTeam()。

解：

修改 FixtureServiceImplUnitTest.java 文件，加入 testListByDateAndTime()、testQueryByField()、testQueryByTeam()三个函数，分别利用 Mockito 和 Assert 语句测试三个接口的返回结果是否符合预期。

6. 使用@ Spring Boot Test (webEnvironment = Spring Boot Test. WebEnvironment. RANDOM)对 FixtureRestController 进行集成测试。

解：

启动集成测试，观察运行结果是否有报错。

7. 阅读 Spring Boot 官方文档，对“院系足球比赛管理系统”完成以下升级：

(1) 定义一个名为 dev 的 profile 作为本机开发环境，并将与本机相关的配置信息放入该 profile 对应的配置文件中，然后基于该 profile 启动 Spring Boot 程序。

（2）通过命令行启动 Spring Boot 项目，通过命令行参数将端口 8091 修改为 9091。

（3）为系统提供的所有 HTTP 接口增加密码保护，通过 HTTP Basic 方式进行认证，用户名为 user，密码为 qlw2c3r4。

（4）为 Spring Boot 项目启用 Actuator 功能，实现对应用运行状态的监控。

解：

新建 application-dev.properties 文件，写入配置信息，例如：

```
server.port=8091#默认端口号
```

使用命令行参数--spring.profiles.active＝dev 和--server.port＝9091 启动 Spring Boot 程序。

编写一个 applicationContext-security.xml 文件：

```
<http pattern="/securityNone" security="none"/>
<http use-expressions="true">
    <intercept-url pattern="/**" access="isAuthenticated()" />
    <http-basic />
</http>

<authentication-manager>
    <authentication-provider>
        <user-service>
            <user name="user" password="q1w2e3r4" authorities="ROLE_USER" />
        </user-service>
    </authentication-provider>
</authentication-manager>
```

在 pom.xml 里加入一条新的 dependency：

```
<dependency>
    <groupId>org.springframework.boot</groupId>
    <artifactId>spring-boot-starter-actuator</artifactId>
</dependency>
```